Systems Biology

Systems Biology

Edited by
Gavin Collingwood

Larsen & Keller
www.larsen-keller.com

Systems Biology
Edited by Gavin Collingwood
ISBN: 978-1-63549-654-3 (Hardback)

▤ Larsen & Keller

Published by Larsen and Keller Education,
5 Penn Plaza,
19th Floor,
New York, NY 10001, USA

Cataloging-in-Publication Data

Systems biology / edited by Gavin Collingwood.
      p. cm.
Includes bibliographical references and index.
ISBN 978-1-63549-654-3
1. Systems biology. 2. Biological systems. 3. Molecular biology.
4. Computational biology. I. Collingwood, Gavin.
QH324.2 .S97 2018
570.28--dc23

For more information regarding Larsen and Keller Education and its products, please visit the publisher's website www.larsen-keller.com

Table of Contents

Preface

Systems biology studies the components of biology. Systems biology has a holistic approach towards the research done on the interactions that occur within biological systems. It integrates the principles from areas such as genomics, biomics, molecular biokinematics, phenomics, transcriptomics, glycomics, lipidomics, etc. As this field is emerging at a rapid pace, the contents of this book will help the readers understand the modern concepts and applications of the subject. It aims to serve as a resource guide for students and experts alike and contribute to the growth of the discipline.

To facilitate a deeper understanding of the contents of this book a short introduction of every chapter is written below:

Chapter 1- Systems biology has a holistic approach in understanding the complexity of the biological system. One of the objectives of the subject is to recognize and discover properties of tissues and organisms by studying gene interactions, biochemical pathways, intracellular mechanics. The chapter on systems biology offers an insightful focus, keeping in mind the complex subject matter.

Chapter 2- The information that is stored in our genes can be transferred into a functional product. One of the examples of these functional products is protein. Some of the techniques of measuring gene expression are reporter gene, western blot, SAGE, DNA microarray and RNA-Seq. The section serves as a source to understand the major categories related to gene expression.

Chapter 3- Mechanisms that enable the production of genetic material are collectively called gene expression regulation. Gene regulator controls exist in a gene regulatory network. The chapter serves as a source to understand the major categories related to gene expression regulation.

Chapter 4- DNA sequences that are close together can be inherited together. This process is known as gene expression. The genetic markers found together have less of a tendency of being separated. The topics discussed in the chapter are of great importance to broaden the existing knowledge on systems biology.

Chapter 5- Enzyme kinetics studies those chemical reactions that are catalyzed by enzymes. Enzymatic reactions speed up the reaction without affecting the reaction equilibrium and hence are important. The aspects elucidated in this chapter are of vital importance, and provide a better understanding of systems biology.

I owe the completion of this book to the never-ending support of my family, who supported me throughout the project.

Editor

A Comprehensive Study of Systems Biology

Systems biology has a holistic approach in understanding the complexity of the biological system. One of the objectives of the subject is to recognize and discover properties of tissues and organisms by studying gene interactions, biochemical pathways, intracellular mechanics. The chapter on systems biology offers an insightful focus, keeping in mind the complex subject matter.

Systems Biology

Systems Biology is an emerging field in Life Sciences which has evinced interest among the community of physicists, chemists, mathematicians and researchers involved in computational modelling to a significant extent. A simple PubMed search would yield more than 25,000 articles in the past two years, with this term appearing either in the title or in the abstract as against three articles in the earlier century. A closer look at the recent articles (which use the term Systems Biology) reveals a systematic work covering computational or experimental efforts to define a system completely. Systems level understanding has dominated themes of discussion in biological sciences for decades. Advances in Molecular Biology, genome level sequencing and interpretations and high throughput measurements at the molecular and imaging level have enabled a new paradigm of understanding a system, its dynamics and the contribution of individual molecules to the performance of a system. Emergence of new and finer techniques in molecular biology enables a systems level understanding grounded with the understanding at the molecular level.

Now, let us align this discussion to a general observation in science. We all know that complex systems are prevalent in the Physics, Engineering, Chemistry and Biology we have discussed over years. The complexity in the system occurs due to a large number of variables in the system, their non linear equations, the kinetic equations and transport equations that describe the system, the network among the variables in the system like Feed Back Loops or gene circuits or networks. Newer techniques and tools in the field have enabled visualization of materials structure and biological molecules at the nano and atomic scale. It is now possible even to determine genome organization and architecture and the molecular players that network to initiate signalling cascades in the cell. Thus one can observe finer kinetic processes, signalling loops and spatio-temporal organization patterns that add to the complexity of the network. Such data sets become substrates on which one can develop theories of complex function and behaviour. This forms the basis of Systems Biology.

Systems biology is the computational and mathematical modeling of complex biological systems. An emerging engineering approach applied to biological scientific research, systems biology is a biology-based interdisciplinary field of study that focuses on complex interactions within biological systems, using a holistic approach (holism instead of the more traditional reductionism) to biological research. Particularly from year 2000 onwards, the concept has been used widely in the biosciences in a variety of contexts. The Human Genome Project is an example of applied systems thinking in biology which has led to new, collaborative ways of working on problems in the biological field of genetics. One of the outreaching aims of systems biology is to model and discover emergent properties, properties of cells, tissues and organisms functioning as a system whose theoretical description is only possible using techniques which fall under the remit of systems biology. These typically involve metabolic networks or cell signaling networks.

Overview

Systems biology can be considered from a number of different aspects:

- As a field of study, particularly, the study of the interactions between the components of biological systems, and how these interactions give rise to the function and behavior of that system (for example, the enzymes and metabolites in a metabolic pathway or the heart beats).

- As a paradigm, usually defined in antithesis to the so-called reductionist paradigm (biological organisation), although fully consistent with the scientific method. The distinction between the two paradigms is referred to in these quotations:

  "The reductionist approach has successfully identified most of the components and many of the interactions but, unfortunately, offers no convincing concepts or methods to understand how system properties emerge...the pluralism of causes and effects in biological networks is better addressed by observing, through quantitative measures, multiple components simultaneously and by rigorous data integration with mathematical models" (Sauer *et al.*).

  "Systems biology...is about putting together rather than taking apart, integration rather than reduction. It requires that we develop ways of thinking about integration that are as rigorous as our reductionist programmes, but different.... It means changing our philosophy, in the full sense of the term" (Denis Noble).

- As a series of operational protocols used for performing research, namely a cycle composed of theory, analytic or computational modelling to propose specific testable hypotheses about a biological system, experimental validation, and then using the newly acquired quantitative description of cells or cell processes to refine the computational model or theory. Since the objective is a model of the interactions in a system, the experimental techniques that most suit systems biology are those that are system-wide and attempt to be as complete as possible. Therefore, transcriptomics, metabolomics, proteomics and high-through-put techniques are used to collect quantitative data for the construction and validation of models.

- As the application of dynamical systems theory to molecular biology. Indeed, the focus on the dynamics of the studied systems is the main conceptual difference between systems biology and bioinformatics.

- As a socioscientific phenomenon defined by the strategy of pursuing integration of complex data about the interactions in biological systems from diverse experimental sources using interdisciplinary tools and personnel.

This variety of viewpoints is illustrative of the fact that systems biology refers to a cluster of peripherally overlapping concepts rather than a single well-delineated field. However the term has widespread currency and popularity as of 2007, with chairs and institutes of systems biology proliferating worldwide.

History

Systems biology finds its roots in:

- the quantitative modeling of enzyme kinetics, a discipline that flourished between 1900 and 1970,

- the mathematical modeling of population dynamics,

- the simulations developed to study neurophysiology, and

- control theory and cybernetics,

- synergetics.

One of the theorists who can be seen as one of the precursors of systems biology is Ludwig von Bertalanffy with his general systems theory. One of the first numerical simulations in cell biology was published in 1952 by the British neurophysiologists and Nobel prize winners Alan Lloyd Hodgkin and Andrew Fielding Huxley, who constructed a mathematical model that explained the action potential propagating along the axon of a neuronal cell. Their model described a cellular function emerging from the interaction between two different molecular components, a potassium and a sodium channel, and can therefore be seen as the beginning of computational systems biology. Also in 1952, Alan Turing published The Chemical Basis of Morphogenesis, describing how non-uniformity could arise in an initially homogeneous biological system.

In 1960, Denis Noble developed the first computer model of the heart pacemaker.

The formal study of systems biology, as a distinct discipline, was launched by systems theorist Mihajlo Mesarovic in 1966 with an international symposium at the Case Institute of Technology in Cleveland, Ohio, entitled "Systems Theory and Biology".

The 1960s and 1970s saw the development of several approaches to study complex molecular systems, such as the metabolic control analysis and the biochemical systems theory. The successes of molecular biology throughout the 1980s, coupled with a skepticism toward theoretical biology, that then promised more than it achieved, caused the quantitative modelling of biological processes to become a somewhat minor field.

However the birth of functional genomics in the 1990s meant that large quantities of high quality data became available, while the computing power exploded, making more realistic models possible. In 1992, then 1994, serial articles on systems medicine, systems genetics and systems biological engineering by B. J. Zeng were published in China, and was giving a lecture on biosystems theory and systems approach research at the First International Conference on Transgenic Animals, Beijing, 1996. In 1997, the group of Masaru Tomita published the first quantitative model of the metabolism of a whole (hypothetical) cell.

Around the year 2000, after Institutes of Systems Biology were established in Seattle and Tokyo, systems biology emerged as a movement in its own right, spurred on by the completion of various genome projects, the large increase in data from the omics (e.g., genomics and proteomics) and the accompanying advances in high-throughput experiments and bioinformatics.

In 2002, the National Science Foundation (NSF) put forward a grand challenge for systems biology in the 21st century to build a mathematical model of the whole cell. In 2003, work at the Massachusetts Institute of Technology was begun on CytoSolve, a method to model the whole cell by dynamically integrating multiple molecular pathway models. Since then, various research institutes dedicated to systems biology have been developed. For example, the NIGMS of NIH established a project grant that is currently supporting over ten systems biology centers in the United States. As of summer 2006, due to a shortage of people in systems biology several doctoral training programs in systems biology have been established in many parts of the world. In that same year, the National Science Foundation (NSF) put forward a grand challenge for systems biology in the 21st century to build a mathematical model of the whole cell. In 2012 the first whole-cell model of Mycoplasma Genitalium was achieved by the Karr Laboratory at the Mount Sinai School of Medicine in New York. The whole-cell model is able to predict viability of M. Genitalium cells in response to genetic mutations.

An important milestone in the development of systems biology has become the international project Physiome.

Associated Disciplines

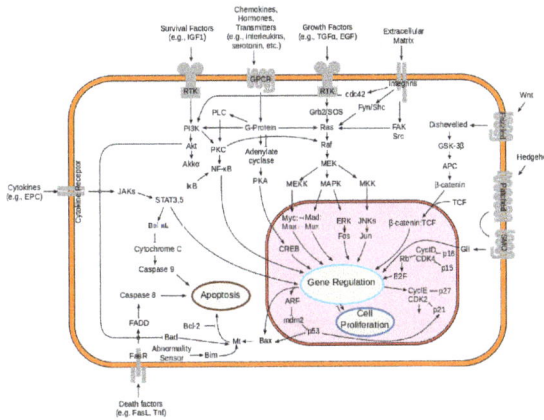

Overview of signal transduction pathways

According to the interpretation of Systems Biology as the ability to obtain, integrate and analyze complex data sets from multiple experimental sources using interdisciplinary tools, some typical technology platforms are:

- Phenomics

  Organismal variation in phenotype as it changes during its life span.

- Genomics

  Organismal deoxyribonucleic acid (DNA) sequence, including intra-organisamal cell specific variation. (i.e., telomere length variation)

- Epigenomics / Epigenetics

  Organismal and corresponding cell specific transcriptomic regulating factors not empirically coded in the genomic sequence. (i.e., DNA methylation, Histone acetylation and deacetylation, etc.).

- Transcriptomics

  Organismal, tissue or whole cell gene expression measurements by DNA microarrays or serial analysis of gene expression

- Interferomics

  Organismal, tissue, or cell-level transcript correcting factors (i.e., RNA interference)

- Proteomics

  Organismal, tissue, or cell level measurements of proteins and peptides via two-dimensional gel electrophoresis, mass spectrometry or multi-dimensional protein identification techniques (advanced HPLC systems coupled with mass spectrometry). Sub disciplines include phosphoproteomics, glycoproteomics and other methods to detect chemically modified proteins.

- Metabolomics

  Organismal, tissue, or cell-level measurements of small molecules known as metabolites

- Glycomics

  Organismal, tissue, or cell-level measurements of carbohydrates

- Lipidomics

  Organismal, tissue, or cell level measurements of lipids.

In addition to the identification and quantification of the above given molecules further techniques analyze the dynamics and interactions within a cell. This includes:

- Interactomics

  Organismal, tissue, or cell level study of interactions between molecules. Currently the authoritative molecular discipline in this field of study is protein-protein interactions (PPI), although the working definition does not preclude inclusion of other molecular disciplines such as those defined here.

- NeuroElectroDynamics

  Organismal, brain computing function as a dynamic system, underlying biophysical mechanisms and emerging computation by electrical interactions.

- Fluxomics

  Organismal, tissue, or cell level measurements of molecular dynamic changes over time.

- Biomics

  Systems analysis of the biome.

- Molecular Biokinematics

  The study of "biology in motion" focused on how cells transit between steady states. Various technologies utilized to capture dynamic changes in mRNA, proteins, and post-translational modifications.

- Semiomics

  Analysis of the system of sign relations of an organism or other biosystem.

- Physiomics

   A systematic study of physiome in biology.

Cancer Systems Biology is an example of the systems biology approach, which can be distinguished by the specific object of study (tumorigenesis and treatment of cancer). It works with the specific data (patient samples, high-throughput data with particular attention to characterizing cancer genome in patient tumour samples) and tools (immortalized cancer cell lines, mouse models of tumorigenesis, xenograft models, Next Generation Sequencing methods, siRNA-based gene knocking down screenings, computational modeling of the consequences of somatic mutations and genome instability). The long-term objective of the systems biology of cancer is ability to better diagnose cancer, classify it and better predict the outcome of a suggested treatment, which is a basis for personalized cancer medicine and virtual cancer patient in more distant prospective. Significant efforts in Computational systems Biology of Cancer have been made in creating realistic multi-scale *in silico* models of various tumours.

The investigations are frequently combined with large-scale perturbation methods, including gene-based (RNAi, mis-expression of wild type and mutant genes) and chemical approaches using small molecule libraries. Robots and automated sensors enable such large-scale experimentation and data acquisition. These technologies are still emerging and many face problems that the larger the quantity of data produced, the lower the quality. A wide variety of quantitative scientists (computational biologists, statisticians, mathematicians, computer scientists, engineers, and physicists) are working to improve the quality of these approaches and to create, refine, and retest the models to accurately reflect observations.

The systems biology approach often involves the development of mechanistic models, such as the reconstruction of dynamic systems from the quantitative properties of their elementary building blocks. For instance, a cellular network can be modelled mathematically using methods coming from chemical kinetics and control theory. Due to the large number of parameters, variables and constraints in cellular networks, numerical and computational techniques are often used (e.g., flux balance analysis).

Bioinformatics and Data Analysis

Other aspects of computer science, informatics, statistics are also used in systems biology. These include:

- New forms of computational model, such as the use of process calculi to model biological processes (notable approaches include stochastic π-calculus, BioAmbients, Beta Binders, BioPEPA, and Brane calculus) and constraint-based modeling.

- Integration of information from the literature, using techniques of information extraction and text mining.

- Development of online databases and repositories for sharing data and models, approaches to database integration and software interoperability via loose coupling of software, websites and databases, or commercial suits.

- Development of syntactically and semantically sound ways of representing biological models.

- Network-based approaches for analyzing high dimensional genomic data sets. For example, weighted correlation network analysis is often used for identifying clusters, modeling the relationship between clusters, calculating fuzzy measures of cluster membership, identifying intramodular hubs, and for studying cluster preservation in other data sets.

- Pathway-based methods for omics data analysis, e.g. approaches to identify and score pathways with differential activity of their gene, protein or metabolite members

Systems biology is aimed at understanding unique biological functions exhibited by biomolecules when they constitute a system as against their behavior as individual macro molecules in isolation. Systems biology concentrates on the non linear interactions between all the components that contribute to the specific biological function of interest. Since a biological system is a complex intricate system, mathematical modelling is required apart from the quantitative experimental assessment of the functioning of the system.

Hence systems biology is a truly interdisciplinary field that brings together mathematics, engineering disciplines such as control systems, signal processing and digital logic into molecular biology to understand bimolecular pathways and interaction networks. The systems paradigm focusses on the complex interactions within biological systems on a holistic approach as against a traditional reductionist approach employed in biological research. This field relies on systems modelling to investigate interesting functional behavior and properties in cells, tissues, genetic systems, cellular reprogramming systems where the theoretical and modelling approach add value to the experimental findings. As said systems biology borrows heavily from inter disciplinary fields such as

- Control theory and Signal Processing.

- Mathematical Modelling of population control

- Quantitative modelling of enzyme kinetics and metabolites in a specific metabolic pathway

- Modelling cell signalling events during development.

Molecular and the Systems Paradigm

As we trace the history of emergence of biology, we recollect the fact that the synthesis of new molecules through biochemistry explained newer reaction pathways and signalling mechanisms in detail. The discovery of the DNA, elucidation of DNA structure and the molecular basis of life, protein isolation techniques, gene identification and functional interpretations-the strong relationship between genetics and biochemistry clearly established that life can be studied at the molecular level. The structural and functional elucidation of newer proteins and their amino acid sequences, fluorescence techniques, advancements in microscopy and a host of biophysical techniques augmented the extent of experimental observations.

Since any living organism is characterized by complex signalling networks and since molecular biology and genetic techniques cannot characterize a living system completely, there are several unknown factors which regulate the dynamics of the system. The advents of large scale genomics and genomic sequencing techniques have today aided gene expression studies at the level of transcriptome, proteome and metabolome. Functional annotation exercises have also enabled our understanding of gene functions.

Systems biology advocates an understanding at the systems level and demands a paradigm shift in' what to look for' in a system. The focus in systems biology is to understand the structure and dynamics of a system under various experimental conditions and under various environments.

- An assembly of genes and proteins cannot be completely interpreted through their network connections though such network mapping might help identify patterns of signal traffic and the control across the cascade.

- Sequencing and functional annotations of all the genes and proteins in an organism will not be sufficient to understand the complexity involved in engineering the organism. It is necessary to understand how the individual components are assembled to execute a typical signalling pathway or a gene expression module in the organism. This is equivalent to constructing an exhaustive gene regulatory network and details of all the biochemical reactions among organisms. This would have limitations in explaining the influence of one domain on the other. But a systems level understanding of the entire biological system will emerge from an insight into the following properties and requires a thorough understanding of genomics, computational science and quantifying techniques as shown in Figure.

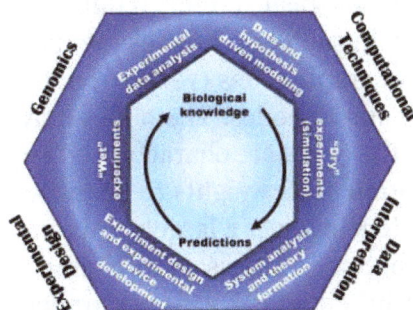

(a): Components of Systems Biology investigation

1) System structures- The gene network interaction and the bio chemical pathways that govern a system and the mechanisms by which these interactions influence the physical properties of intra cellular and multi cellular components.

2) System dynamics- Metabolic analysis, sensitivity analysis and dynamic analysis to understand how the behavior of the system changes over time.

3) Control methods- Mechanisms that regulate the state of the cell. This information can be modulated.

4) Design method- Strategically designed and engineered biological systems with desired properties.

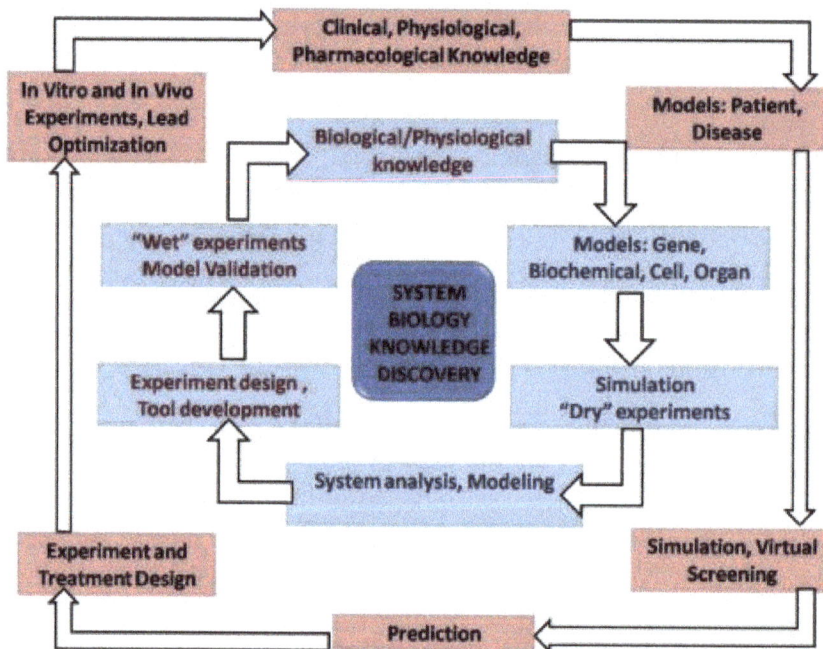

(b) A Typical Systems Biology Paradigm

If we consider biological systems, we note that some of the systems show linear chain of interactions which means that the network of interactions in the system is organized and straight forward. Certain biological systems require a formal representation of the networks using graph theory. In such cases the elements of the system are reduced to the simplest representation of graph nodes (vertices) and their pair wise relationships as edges (links) connecting pairs of nodes. The nodes may be genes or mRNA or proteins or other biomolecules. The edges are characterized by positive signs for activation and negative signs for inhibition or as weights to quantify reaction rates, strengths or confidence levels. Since a system involves a large number of cellular components which show temporal variation, it becomes necessary to characterize the nodes by quantitative information which describes the mRNA copy number or the concentration of the molecules involved in the reaction.

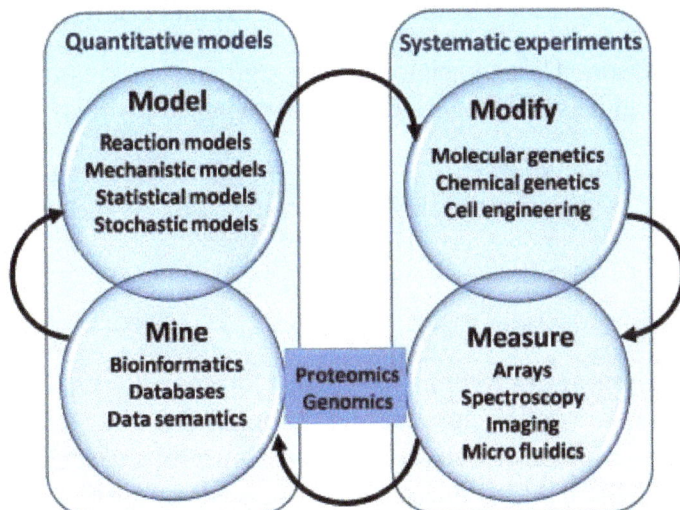

(c) Modify-measure-mine-model paradigm

Systems level analysis of a regulatory system demands accurate measurements and high throughput experimental measurements. Newer technologies like single molecule techniques, micro fluidic systems, fluorescence correlation spectroscopy enable minute rapid and precise measurements of the system. Systems Biology Markup language, Systems Biology work bench are some of the modular open source softwares for research in this exciting field.

Understanding Biology at the Systems Level

What does a 'systems level understanding' actually mean? Molecular Biology techniques focus on biomolecules - molecules such as nucleic acids and proteins, their sequence, structural and function inside the cell. Systems Biology tries to understand a system which is built of these molecular components. If one wants to measure the interactions in a system, one notices that the interactions vary with respect to molecules; the molecules exhibit properties on their own as well as the properties they acquire through interactions. Thus a biological function emerges because of the collective behaviour of the molecules in the system and not because of molecules in isolation. The structure of the system, the network topologies, the functionality of the constituent components and their dynamics constitute the synergistic system. In a simple language this means that systems Biology attempts to understand the structure and dynamics of a system as a whole rather than understanding its genes and proteins in isolation. The following properties provide a systems level understanding of a biological system.

1. Systems Structure-Gene interactions, biochemical pathways, intracellular mechanisms

2. Systems Dynamics- Metabolic, sensitivity and dynamic analysis to study system behaviour

3. Control Method- Modulation of mechanisms that regulate cell states

4. Design Method- Using basic principles and simulations to modify and construct biological systems with desired properties.

Systems Biology does not confine itself to a single organism or single set of macro molecules or to specific techniques. It tries to integrate all these and it is rather heterogeneous.

What does one Require for a Systems Level Understanding?

To carry out an analysis at the systems level, one requires a comprehensive set of quantitative data. Such quantitative data can be obtained from experiments, genome sequencing efforts and molecular network data or from projects such as the Alliance for Cellular Signalling (AfCS) which encompass large scale quantification of cellular data with the sole aim of simulating cellular models. A systems level analysis involving modelling should be explored at the preliminary stages of the project. This would facilitate identification of bottle necks in measuring certain quantities required for building the final model and also to overcome difficulties due to insufficient and inaccurate data which will not add value for model building.

Measurements should be precise and comprehensive and demand the following factors for precise definition.

1. Factor Comprehensiveness

eg. the number of mRNA transcripts or proteins that can be measured simultaneously.

2. Temporal Comprehensiveness- The time scales in which such minute measurements are made.

3. Parameter Comprehensiveness- Simultaneous measurements of functional parameters inside the cell such as protein phosphorylation, localization, mRNA concentration, etc. Experiments planned with a modelling approach can clearly predict areas where accuracy is most essential and areas where it is not so important. This approach helps in optimal allocation of resources.

Systems Biology therefore facilitates understanding of

- Biological structure as well as network architecture of the system.

- Qualitative and quantitative dynamics of the system supported by predicted modelling

- Control points in the system

- Design methodologies for the system.

This integration of modelling with quantitative experimentation helps us to obtain information not yielded by experimental approaches alone. Systems Biology thus generates key hypothesis that augments our understanding of the complex interactions or functions of biological systems. An often quoted example is the phenomenon of calcium signal transduction in which calcium oscillations where predicted through computational modelling. Calcium dynamics on CaMPK 2, the stochasticity on single calcium channels influencing the dynamics of the system have been investigated in detail through experimental and modelling approaches.

Computing and Systems Biology

Integration of computing into the analysis of the biological system leads to a better understanding of the functional components of the system. The Systems Biology Markup Language (SBML) is an elegant format to represent models of biological processes. One can simulate metabolic reaction, cell signalling systems and logical function exhibited by the biological system. SBML2LATEX converts SBML files into LATEX files. MAT-LAB, the versatile tool for engineering calculations has integrated a SBML tool box built on top of libSBML. This tool facilitates the use of SBMLs in MATLAB. Molecular interaction networks, gene expression profiles, protein-protein interactions, can be visualized using "Cytoscape", an open source tool which facilitates plug-in development using its open API based on JAVA. We have "hands –on" modules on these tools as we learn the essentials of Systems Biology.

Modelling Biological Systems

Modelling biological systems is a significant task of systems biology and mathematical biology. Computational systems biology aims to develop and use efficient algorithms, data structures, visualization and communication tools with the goal of computer modelling of biological systems. It involves the use of computer simulations of biological systems, including cellular subsystems (such as the networks of metabolites and enzymes which comprise metabolism, signal transduction pathways and gene regulatory networks), to both analyze and visualize the complex connections of these cellular processes.

Artificial life or virtual evolution attempts to understand evolutionary processes via the computer simulation of simple (artificial) life forms.

Overview

It is understood that an unexpected emergent property of a complex system is a result of the interplay of the cause-and-effect among simpler, integrated parts. Biological systems manifest many important examples of emergent properties in the complex interplay of components. Traditional study of biological systems requires reductive methods in which quantities of data are gathered by category, such as concentration over time in response to a certain stimulus. Computers are critical to analysis and modelling of these data. The goal is to create accurate real-time models of a system's response to environmental and internal stimuli, such as a model of a cancer cell in order to find weaknesses in its signalling pathways, or modelling of ion channel mutations to see effects on cardiomyocytes and in turn, the function of a beating heart.

Standards

By far the most widely accepted standard format for storing and exchanging models in the field is the Systems Biology Markup Language (SBML) The SBML.org website includes a guide to many important software packages used in computational systems biology. Other markup languages with different emphases include BioPAX and CellML.

Particular Tasks

Cellular Model

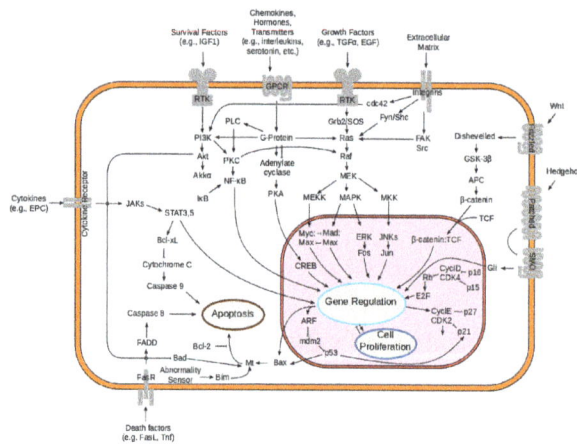

Part of the Cell Cycle

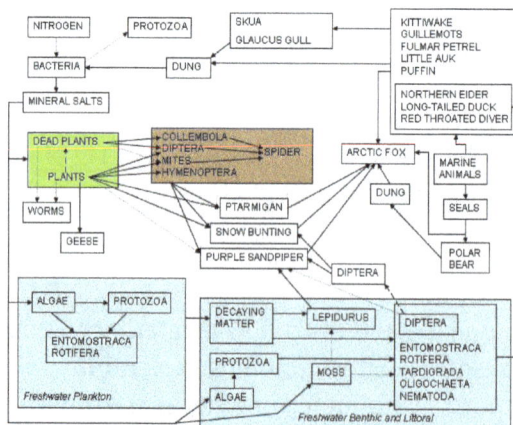

Summerhayes and Elton's 1923 food web of Bear Island (*Arrows represent an organism being consumed by another organism*).

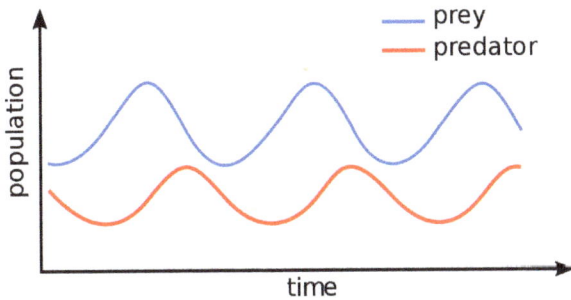

A sample time-series of the Lotka–Volterra model. Note that the two
populations exhibit cyclic behaviour.

Creating a cellular model has been a particularly challenging task of systems biology
and mathematical biology. It involves the use of computer simulations of the many
cellular subsystems such as the networks of metabolites and enzymes which comprise
metabolism, signal transduction pathways and gene regulatory networks to both analyze and visualize the complex connections of these cellular processes.

The complex network of biochemical reaction/transport processes and their spatial
organization make the development of a predictive model of a living cell a grand challenge for the 21st century, listed as such by the National Science Foundation (NSF)
in 2006.

A whole cell computational model for the bacterium *Mycoplasma genitalium*, includ-
ing all its 525 genes, gene products, and their interactions, was built by scientists from
Stanford University and the J. Craig Venter Institute and published on 20 July 2012 in
Cell.

A dynamic computer model of intracellular signaling was the basis for Merrimack
Pharmaceuticals to discover the target for their cancer medicine MM-111.

Membrane computing is the task of modelling specifically a cell membrane.

Multi-cellular Organism Simulation

An open source simulation of C. elegans at the cellular level is being pursued by the
OpenWorm community. So far the physics engine Gepetto has been built and models
of the neural connectome and a muscle cell have been created in the NeuroML format.

Protein Folding

Protein structure prediction is the prediction of the three-dimensional structure of
a protein from its amino acid sequence—that is, the prediction of a protein's tertiary
structure from its primary structure. It is one of the most important goals pursued by
bioinformatics and theoretical chemistry. Protein structure prediction is of high importance in medicine (for example, in drug design) and biotechnology (for example, in

the design of novel enzymes). Every two years, the performance of current methods is assessed in the CASP experiment.

Human Biological Systems

Brain Model

The Blue Brain Project is an attempt to create a synthetic brain by reverse-engineering the mammalian brain down to the molecular level. The aim of this project, founded in May 2005 by the Brain and Mind Institute of the *École Polytechnique* in Lausanne, Switzerland, is to study the brain's architectural and functional principles. The project is headed by the Institute's director, Henry Markram. Using a Blue Gene supercomputer running Michael Hines's NEURON software, the simulation does not consist simply of an artificial neural network, but involves a partially biologically realistic model of neurons. It is hoped by its proponents that it will eventually shed light on the nature of consciousness. There are a number of sub-projects, including the Cajal Blue Brain, coordinated by the Supercomputing and Visualization Center of Madrid (CeSViMa), and others run by universities and independent laboratories in the UK, U.S., and Israel. The Human Brain Project builds on the work of the Blue Brain Project. It is one of six pilot projects in the Future Emerging Technologies Research Program of the European Commission, competing for a billion euro funding.

Model of the Immune System

The last decade has seen the emergence of a growing number of simulations of the immune system.

Virtual Liver

The Virtual Liver project is a 43 million euro research program funded by the German Government, made up of seventy research group distributed across Germany. The goal is to produce a virtual liver, a dynamic mathematical model that represents human liver physiology, morphology and function.

Tree Model

Electronic trees (e-trees) usually use L-systems to simulate growth. L-systems are very important in the field of complexity science and A-life. A universally accepted system for describing changes in plant morphology at the cellular or modular level has yet to be devised. The most widely implemented tree generating algorithms are described in the papers "Creation and Rendering of Realistic Trees", and Real-Time Tree Rendering

Ecological Models

Ecosystem models are mathematical representations of ecosystems. Typically they simplify complex foodwebs down to their major components or trophic levels, and quantify these as either numbers of organisms, biomass or the inventory/concentration of some pertinent chemical element (for instance, carbon or a nutrient species such as nitrogen or phosphorus).

Models in Ecotoxicology

The purpose of models in ecotoxicology is the understanding, simulation and prediction of effects caused by toxicants in the environment. Most current models describe effects on one of many different levels of biological organization (e.g. organisms or populations). A challenge is the development of models that predict effects across biological scales. Ecotoxicology and models discusses some types of ecotoxicological models and provides links to many others.

Modelling of Infectious Disease

It is possible to model the progress of most infectious diseases mathematically to discover the likely outcome of an epidemic or to help manage them by vaccination. This field tries to find parameters for various infectious diseases and to use those parameters to make useful calculations about the effects of a mass vaccination programme.

Bioinformatics

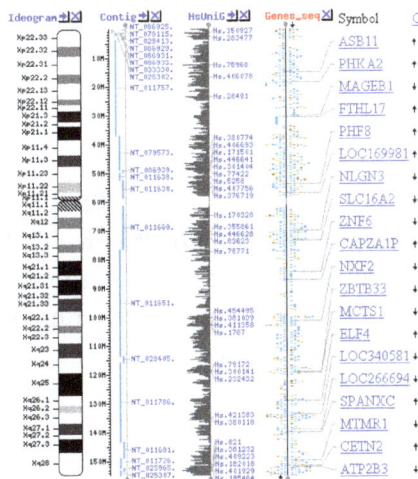

Map of the human X chromosome (from the National Center for Biotechnology Information website).

Bioinformatics is an interdisciplinary field that develops methods and software tools for understanding biological data. As an interdisciplinary field of science, bioinformatics combines computer science, statistics, mathematics, and engineering to analyze and interpret biological data. Bioinformatics has been used for *in silico* analyses of biological queries using mathematical and statistical techniques.

Bioinformatics is both an umbrella term for the body of biological studies that use computer programming as part of their methodology, as well as a reference to specific analysis "pipelines" that are repeatedly used, particularly in the field of genomics. Common uses of bioinformatics include the identification of candidate genes and nucleotides (SNPs). Often, such identification is made with the aim of better understanding the genetic basis of disease, unique adaptations, desirable properties (esp. in agricultural species), or differences between populations. In a less formal way, bioinformatics also tries to understand the organisational principles within nucleic acid and protein sequences, called proteomics.

Introduction

Bioinformatics has become an important part of many areas of biology. In experimental molecular biology, bioinformatics techniques such as image and signal processing allow extraction of useful results from large amounts of raw data. In the field of genetics and genomics, it aids in sequencing and annotating genomes and their observed mutations. It plays a role in the text mining of biological literature and the development of biological and gene ontologies to organize and query biological data. It also plays a role in the analysis of gene and protein expression and regulation. Bioinformatics tools aid in the comparison of genetic and genomic data and more generally in the understanding of evolutionary aspects of molecular biology. At a more integrative level, it helps analyze and catalogue the biological pathways and networks that are an important part of systems biology. In structural biology, it aids in the simulation and modeling of DNA, RNA, proteins as well as biomolecular interactions.

History

Historically, the term *bioinformatics* did not mean what it means today. Paulien Hogeweg and Ben Hesper coined it in 1970 to refer to the study of information processes in biotic systems. This definition placed bioinformatics as a field parallel to biophysics (the study of physical processes in biological systems) or biochemistry (the study of chemical processes in biological systems).

Sequences

5'ATGACGTGGGGA3'
3'TACTGCACCCCT5'

Sequences of genetic material are frequently used in bioinformatics
and are easier to manage using computers than manually.

Computers became essential in molecular biology when protein sequences became available after Frederick Sanger determined the sequence of insulin in the early 1950s. Comparing multiple sequences manually turned out to be impractical. A pioneer in the field was Margaret Oakley Dayhoff, who has been hailed by David Lipman, director of the National Center for Biotechnology Information, as the "mother and father of bioinformatics." Dayhoff compiled one of the first protein sequence databases, initially published as books and pioneered methods of sequence alignment and molecular evolution. Another early contributor to bioinformatics was Elvin A. Kabat, who pioneered biological sequence analysis in 1970 with his comprehensive volumes of antibody sequences released with Tai Te Wu between 1980 and 1991.

Goals

To study how normal cellular activities are altered in different disease states, the biological data must be combined to form a comprehensive picture of these activities. Therefore, the field of bioinformatics has evolved such that the most pressing task now involves the analysis and interpretation of various types of data. This includes nucleotide and amino acid sequences, protein domains, and protein structures. The actual process of analyzing and interpreting data is referred to as computational biology. Important sub-disciplines within bioinformatics and computational biology include:

- Development and implementation of computer programs that enable efficient access to, use and management of, various types of information

- Development of new algorithms (mathematical formulas) and statistical measures that assess relationships among members of large data sets. For example, there are methods to locate a gene within a sequence, to predict protein structure and/or function, and to cluster protein sequences into families of related sequences.

The primary goal of bioinformatics is to increase the understanding of biological processes. What sets it apart from other approaches, however, is its focus on developing and applying computationally intensive techniques to achieve this goal. Examples include: pattern recognition, data mining, machine learning algorithms, and visualization. Major research efforts in the field include sequence alignment, gene finding, genome assembly, drug design, drug discovery, protein structure alignment, protein structure prediction, prediction of gene expression and protein–protein interactions, genome-wide association studies, the modeling of evolution and cell division/mitosis.

Bioinformatics now entails the creation and advancement of databases, algorithms, computational and statistical techniques, and theory to solve formal and practical problems arising from the management and analysis of biological data.

Over the past few decades, rapid developments in genomic and other molecular research technologies and developments in information technologies have combined to produce a tremendous amount of information related to molecular biology. Bioinformatics is the name given to these mathematical and computing approaches used to glean understanding of biological processes.

Common activities in bioinformatics include mapping and analyzing DNA and protein sequences, aligning DNA and protein sequences to compare them, and creating and viewing 3-D models of protein structures.

Relation to other Fields

Bioinformatics is a science field that is similar to but distinct from biological computation, while it is often considered synonymous to computational biology. Biological computation uses bioengineering and biology to build biological computers, whereas bioinformatics uses computation to better understand biology. Bioinformatics and computational biology involve the analysis of biological data, particularly DNA, RNA, and protein sequences. The field of bioinformatics experienced explosive growth starting in the mid-1990s, driven largely by the Human Genome Project and by rapid advances in DNA sequencing technology.

Analyzing biological data to produce meaningful information involves writing and running software programs that use algorithms from graph theory, artificial intelligence, soft computing, data mining, image processing, and computer simulation. The algorithms in turn depend on theoretical foundations such as discrete mathematics, control theory, system theory, information theory, and statistics.

Sequence Analysis

```
A5ASC3.1    14 SIKLWPPSQTTRLLLVERMANNLST..PSIFTRK..YGSLSKEEARENAKQIEEVACSTANQ.....HYEKEPDGDGGSAVQLYAKECSKLILEVLK 101
B4F917.1    13 SIKLWPPSESTRIMLVDRMTNNLST...ESIFSRK..YRLLGKQEAHENAKTIEELCFALADE.....HFREEPDGDGSSAVQLYAKETSKMMLEVLK 100
A9S1V2.1    23 VFKLWPPSQGTREAVRQKMALKLSS..ACFESQS..FARIELADAQEHARAIEEVAFGAAQE......ADSGGDKTGSAVVMVYAKHASKLMLETLR 109
B9GSN7.1    13 SVKLWPPGQSTRLMLVERMTKNFIT..PSFISRK..YGLLSKEEAEEDAKKIEEVAFAAANQ.....HYEKQPDGDGSSAVQIYAKESSRLMLEVLK 100
Q8HO56.1    30 SFSIWPPTQRTRDAVVRRLVDTLGG..DTILCKR..YGAVPAADAEPAARGIEAEAFDAAAA..SGEAAATASVEEGIKALQLYSKEVSRRLLDFVK 120
Q0D4Z3.2    44 SLSIWPPSQRTRDAVVRRLVQTLVA..PSILSQR..YGAVPEAEAGRAAAAVEAEAYAAVTES.SSAAAAPASVEDGIEVLQAYSKEVSRRLLELAK 135
B9MVW8.1    56 SFSIWPPTQRTRDAIISRLIETLST..TSVLSKR..YGTIPKEEASEASRRIEEEAFSGAST.......VASSEKDGLEVLQLYSKEISKRMLETVK 141
Q0IYC5.1    29 SFAVWPPTRRTRDAVVRRLVAVLSGDTTTALRKKRYRYGAVPAADAERAARAVEAQAFDAASA...SSSSSSSVEDGIETLQLYSREVSNRLLAFVR 121
A9NWJ46.1   13 SIKLWPPSESTRLMLVERMTDNLSS..VSFFSRK..YGLLSKEEAAENAKRIEETAFLAAND.....HEAKEPNLDDSSVVQFYAREASKLMLEALK 100
Q9C500.1    57 SLRIWPPTQKTRDAVLNRLIETLST..ESILSKR..YGTLKSDDATTVAKLIEEEAYGVASN......AVSSDDDGIKILELYSKEISKRMLESVK 142
Q2HRI7.1    25 MYSIWPPKQRTRDAVKNRLIETLST..PSVLTKR..YGTMSADEASAAAIQIEDEAFSVANA.......SSSTSNDNVTILEVYSKEISKRMIETVK 110
Q9M7N3.1    28 SFKIWPPTQRTREAVVRRLVETLTS..QSVLSKR..YGVIPEEDATSAARIEEEAFSVASV.ASAASTGGRPEDEWIEVLHIYSQEIXQRVVESAK 119
Q9M7N6.1    25 SFSIWPPTQRTRDAVINRLIESLST..PSILSKR..YGTLPQDEASETARLIEEEAFAAAGS.......TASDADDGIEILQVYSKEISKRMIDTVK 110
Q9LE82.1    14 SVKMWPPSKSTRLMLVERMTKNITT..PSIFSRK..YGLLSVEEAEQDAKRIEDLAFATANK.....HFQNEPDGDGTSAVHVYAKESSKLMLDVIK 101
Q9M651.2    13 SIKLWPPSLPTRKALIERITNNFSS..KTIFTEK..YGSLTKDQATENAKRIEDIAFSTANQ.....QFEREPDGDGGSAVQLYAKECSKLILEVLK 100
B9R748.1    48 SLSIWPPTQRTRDAVITRLIETLSS..PSVLSKR..YGTISHDEAESAARRIEDEAFGVANT.......ATSAEDDGLEILQLYSKEISRRMLDTVK 133
```

The sequences of different genes or proteins may be aligned side-by-side to measure their similarity. This alignment compares protein sequences containing WPP domains.

Since the Phage Φ-X174 was sequenced in 1977, the DNA sequences of thousands of organisms have been decoded and stored in databases. This sequence information is analyzed to determine genes that encode proteins, RNA genes, regulatory sequences, structural motifs, and repetitive sequences. A comparison of genes within a species or between different species can show similarities between protein functions, or relations between species (the use of molecular systematics to construct phylogenetic trees). With the growing amount of data, it long ago became impractical to analyze DNA sequences manually. Today, computer programs such as BLAST are used daily to search sequences from more than 260 000 organisms, containing over 190 billion nucleotides.

These programs can compensate for mutations (exchanged, deleted or inserted bases) in the DNA sequence, to identify sequences that are related, but not identical. A variant of this sequence alignment is used in the sequencing process itself.

DNA Sequencing

Before sequences can be analyzed they have to be obtained. DNA sequencing is still a non-trivial problem as the raw data may be noisy or afflicted by weak signals. Algorithms have been developed for base calling for the various experimental approaches to DNA sequencing.

Sequence Assembly

Most DNA sequencing techniques produce short fragments of sequence that need to be assembled to obtain complete gene or genome sequences. The so-called shotgun sequencing technique (which was used, for example, by The Institute for Genomic Research (TIGR) to sequence the first bacterial genome, *Haemophilus influenzae*) generates the sequences of many thousands of small DNA fragments (ranging from 35 to 900 nucleotides long, depending on the sequencing technology). The ends of these fragments overlap and, when aligned properly by a genome assembly program, can be used to reconstruct the complete genome. Shotgun sequencing yields sequence data quickly, but the task of assembling the fragments can be quite complicated for larger genomes. For a genome as large as the human genome, it may take many days of CPU time on large-memory, multiprocessor computers to assemble the fragments, and the resulting assembly usually contains numerous gaps that must be filled in later. Shotgun sequencing is the method of choice for virtually all genomes sequenced today, and genome assembly algorithms are a critical area of bioinformatics research.

Genome Annotation

In the context of genomics, annotation is the process of marking the genes and other biological features in a DNA sequence. This process needs to be automated because most genomes are too large to annotate by hand, not to mention the desire to annotate as many genomes as possible, as the rate of sequencing has ceased to pose a bottleneck. Annotation is made possible by the fact that genes have recognisable start and stop regions, although the exact sequence found in these regions can vary between genes.

The first genome annotation software system was designed in 1995 by Owen White, who was part of the team at The Institute for Genomic Research that sequenced and analyzed the first genome of a free-living organism to be decoded, the bacterium *Haemophilus influenzae*. White built a software system to find the genes (fragments of genomic sequence that encode proteins), the transfer RNAs, and to make initial assignments of function to those genes. Most current genome annotation systems work similarly, but the programs available for analysis of genomic DNA, such as the GeneMark program trained and used to find protein-coding genes in *Haemophilus influenzae*, are constantly changing and improving.

Following the goals that the Human Genome Project left to achieve after its closure in 2003, a new project developed by the National Human Genome Research Institute in the U.S appeared. The so-called ENCODE project is a collaborative data collection of the functional elements of the human genome that uses next-generation DNA-sequencing technologies and genomic tiling arrays, technologies able to generate automatically large amounts of data with lower research costs but with the same quality and viability.

Computational Evolutionary Biology

Evolutionary biology is the study of the origin and descent of species, as well as their change over time. Informatics has assisted evolutionary biologists by enabling researchers to:

- trace the evolution of a large number of organisms by measuring changes in their DNA, rather than through physical taxonomy or physiological observations alone,

- more recently, compare entire genomes, which permits the study of more complex evolutionary events, such as gene duplication, horizontal gene transfer, and the prediction of factors important in bacterial speciation,

- build complex computational population genetics models to predict the outcome of the system over time

- track and share information on an increasingly large number of species and organisms

Future work endeavours to reconstruct the now more complex tree of life.

The area of research within computer science that uses genetic algorithms is sometimes confused with computational evolutionary biology, but the two areas are not necessarily related.

Comparative Genomics

The core of comparative genome analysis is the establishment of the correspondence between genes (orthology analysis) or other genomic features in different organisms. It is these intergenomic maps that make it possible to trace the evolutionary processes responsible for the divergence of two genomes. A multitude of evolutionary events acting at various organizational levels shape genome evolution. At the lowest level, point mutations affect individual nucleotides. At a higher level, large chromosomal segments undergo duplication, lateral transfer, inversion, transposition, deletion and insertion. Ultimately, whole genomes are involved in processes of hybridization, polyploidization and endosymbiosis, often leading to rapid speciation. The complexity of genome evolution poses many exciting challenges to developers of mathematical models and algorithms, who have recourse to a spectra of algorithmic, statistical and mathematical techniques, ranging from exact, heuristics, fixed parameter and approximation algorithms for problems based on parsimony models to Markov Chain Monte Carlo algorithms for Bayesian analysis of problems based on probabilistic models.

Many of these studies are based on the homology detection and protein families computation.

Pan Genomics

Pan genomics is a concept introduced in 2005 by Tettelin and Medini which eventually took root in bioinformatics. Pan genome is the complete gene repertoire of a particular taxonomic group: although initially applied to closely related strains of a species, it can be applied to a larger context like genus, phylum etc. It is divided in two parts- The Core genome: Set of genes common to all the genomes under study (These are often housekeeping genes vital for survival) and The Dispensable/Flexible Genome: Set of genes not present in all but one or some genomes under study. a bioinformatics tool BPGA can be used to characterize the Pan Genome of bacterial species.

Genetics of Disease

With the advent of next-generation sequencing we are obtaining enough sequence data to map the genes of complex diseases such as diabetes, infertility, breast cancer or Alzheimer's Disease. Genome-wide association studies are a useful approach to pinpoint the mutations responsible for such complex diseases. Through these studies, thousands of DNA variants have been identified that are associated with similar diseases and traits. Furthermore, the possibility for genes to be used at prognosis, diagnosis or treatment is one of the most essential applications. Many studies are discussing both the promising ways to choose the genes to be used and the problems and pitfalls of using genes to predict disease presence or prognosis.

Analysis of Mutations in Cancer

In cancer, the genomes of affected cells are rearranged in complex or even unpredictable ways. Massive sequencing efforts are used to identify previously unknown point mutations in a variety of genes in cancer. Bioinformaticians continue to produce specialized automated systems to manage the sheer volume of sequence data produced, and they create new algorithms and software to compare the sequencing results to the growing collection of human genome sequences and germline polymorphisms. New physical detection technologies are employed, such as oligonucleotide microarrays to identify chromosomal gains and losses (called comparative genomic hybridization), and single-nucleotide polymorphism arrays to detect known *point mutations*. These detection methods simultaneously measure several hundred thousand sites throughout the genome, and when used in high-throughput to measure thousands of samples, generate terabytes of data per experiment. Again the massive amounts and new types of data generate new opportunities for bioinformaticians. The data is often found to contain considerable variability, or noise, and thus Hidden Markov model and change-point analysis methods are being developed to infer real copy number changes.

With the breakthroughs that this next-generation sequencing technology is providing to the field of Bioinformatics, cancer genomics could drastically change. These new methods and software allow bioinformaticians to sequence many cancer genomes quickly and affordably. This could create a more flexible process for classifying types of cancer by analysis of cancer driven mutations in the genome. Furthermore, tracking of patients while the disease progresses may be possible in the future with the sequence of cancer samples.

Another type of data that requires novel informatics development is the analysis of lesions found to be recurrent among many tumors.

Gene and Protein Expression

Analysis of Gene Expression

The expression of many genes can be determined by measuring mRNA levels with multiple techniques including microarrays, expressed cDNA sequence tag (EST) sequencing, serial analysis of gene expression (SAGE) tag sequencing, massively parallel signature sequencing (MPSS), RNA-Seq, also known as "Whole Transcriptome Shotgun Sequencing" (WTSS), or various applications of multiplexed in-situ hybridization. All of these techniques are extremely noise-prone and/or subject to bias in the biological measurement, and a major research area in computational biology involves developing statistical tools to separate signal from noise in high-throughput gene expression studies. Such studies are often used to determine the genes implicated in a disorder: one might compare microarray data from cancerous epithelial cells to data from non-cancerous cells to determine the transcripts that are up-regulated and down-regulated in a particular population of cancer cells.

Analysis of Protein Expression

Protein microarrays and high throughput (HT) mass spectrometry (MS) can provide a snapshot of the proteins present in a biological sample. Bioinformatics is very much involved in making sense of protein microarray and HT MS data; the former approach faces similar problems as with microarrays targeted at mRNA, the latter involves the problem of matching large amounts of mass data against predicted masses from protein sequence databases, and the complicated statistical analysis of samples where multiple, but incomplete peptides from each protein are detected.

Analysis of Regulation

Regulation is the complex orchestration of events by which a signal, potentially an extracellular signal such as a hormone, eventually leads to an increase or decrease in the activity of one or more proteins. Bioinformatics techniques have been applied to explore various steps in this process.

For example, gene expression can be regulated by nearby elements in the genome. Promoter analysis involves the identification and study of sequence motifs in the DNA surrounding the coding region of a gene. These motifs influence the extent to which that region is transcribed into mRNA. Enhancer elements far away from the promoter can also regulate gene expression, through three-dimensional looping interactions. These interactions can be determined by bioinformatic analysis of chromosome conformation capture experiments.

Expression data can be used to infer gene regulation: one might compare microarray data from a wide variety of states of an organism to form hypotheses about the genes involved in each state. In a single-cell organism, one might compare stages of the cell cycle, along with various stress conditions (heat shock, starvation, etc.). One can then apply clustering algorithms to that expression data to determine which genes are co-expressed. For example, the upstream regions (promoters) of co-expressed genes can be searched for over-represented regulatory elements. Examples of clustering algorithms applied in gene clustering are k-means clustering, self-organizing maps (SOMs), hierarchical clustering, and consensus clustering methods.

Analysis of Cellular Organization

Several approaches have been developed to analyze the location of organelles, genes, proteins, and other components within cells. This is relevant as the location of these components affects the events within a cell and thus helps us to predict the behavior of biological systems. A gene ontology category, *cellular compartment*, has been devised to capture subcellular localization in many biological databases.

Microscopy and Image Analysis

Microscopic pictures allow us to locate both organelles as well as molecules. It may also help us to distinguish between normal and abnormal cells, e.g. in cancer.

Protein Localization

The localization of proteins helps us to evaluate the role of a protein. For instance, if a protein is found in the nucleus it may be involved in gene regulation or splicing. By contrast, if a protein is found in mitochondria, it may be involved in respiration or other metabolic processes. Protein localization is thus an important component of protein function prediction. There are well developed protein subcellular localization prediction resources available, including protein subcellualr location databases and prediction tools.

Nuclear Organisation of Chromatin

Data from high-throughput chromosome conformation capture experiments, such as Hi-C (experiment) and ChIA-PET, can provide information on the spatial proximity of DNA loci. Analysis of these experiments can determine the three-dimensional structure and nuclear organization of chromatin. Bioinformatic challenges in this field include partitioning the genome into domains, such as Topologically Associating Domains (TADs), that are organised together in three-dimensional space.

Structural Bioinformatics

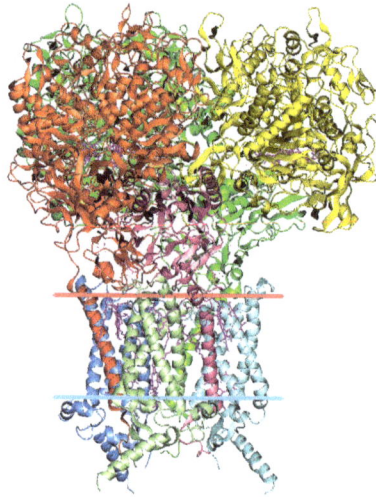

3-dimensional protein structures such as this one
are common subjects in bioinformatic analyses.

Protein structure prediction is another important application of bioinformatics. The amino acid sequence of a protein, the so-called primary structure, can be easily determined from the sequence on the gene that codes for it. In the vast majority of cases, this primary structure uniquely determines a structure in its native environment. (Of course, there are exceptions, such as the bovine spongiform encephalopathy – a.k.a. Mad Cow Disease – prion.) Knowledge of this structure is vital in understanding the function of the protein. Structural information is usually classified as one of *secondary*, *tertiary* and *quaternary* structure. A viable general solution to such predictions remains an open problem. Most efforts have so far been directed towards heuristics that work most of the time.

One of the key ideas in bioinformatics is the notion of homology. In the genomic branch of bioinformatics, homology is used to predict the function of a gene: if the sequence of gene *A*, whose function is known, is homologous to the sequence of gene *B*, whose function is unknown, one could infer that B may share A's function. In the structural branch of bioinformatics, homology is used to determine which parts of a protein are important in structure formation and interaction with other proteins. In a technique called homology modeling, this information is used to predict the structure of a protein once the structure of a homologous protein is known. This currently remains the only way to predict protein structures reliably.

One example of this is the similar protein homology between hemoglobin in humans and the hemoglobin in legumes (leghemoglobin). Both serve the same purpose of transporting oxygen in the organism. Though both of these proteins have completely different amino acid sequences, their protein structures are virtually identical, which reflects their near identical purposes.

Other techniques for predicting protein structure include protein threading and *de novo* (from scratch) physics-based modeling.

Network and Systems Biology

Network analysis seeks to understand the relationships within biological networks such as metabolic or protein–protein interaction networks. Although biological networks can be constructed from a single type of molecule or entity (such as genes), network biology often attempts to integrate many different data types, such as proteins, small molecules, gene expression data, and others, which are all connected physically, functionally, or both.

Systems biology involves the use of computer simulations of cellular subsystems (such as the networks of metabolites and enzymes that comprise metabolism, signal transduction pathways and gene regulatory networks) to both analyze and visualize the complex connections of these cellular processes. Artificial life or virtual evolution attempts to understand evolutionary processes via the computer simulation of simple (artificial) life forms.

Molecular Interaction Networks

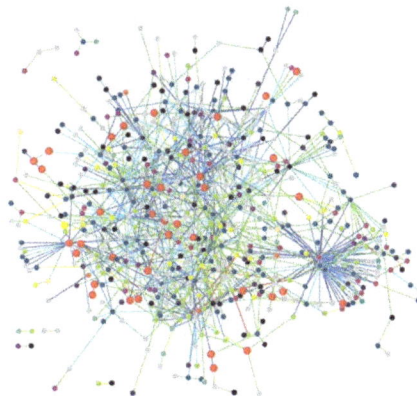

Interactions between proteins are frequently visualized and analyzed using networks. This network is made up of protein–protein interactions from *Treponema pallidum*, the causative agent of syphilis and other diseases.

Other interactions encountered in the field include Protein–ligand (including drug) and protein–peptide. Molecular dynamic simulation of movement of atoms about rotatable bonds is the fundamental principle behind computational algorithms, termed docking algorithms, for studying molecular interactions.

Others

Literature Analysis

The growth in the number of published literature makes it virtually impossible to read every paper, resulting in disjointed sub-fields of research. Literature analysis aims to employ computational and statistical linguistics to mine this growing library of text resources. For example:

- Abbreviation recognition – identify the long-form and abbreviation of biological terms

- Named entity recognition – recognizing biological terms such as gene names

- Protein–protein interaction – identify which proteins interact with which proteins from text

The area of research draws from statistics and computational linguistics.

High-throughput Image Analysis

Computational technologies are used to accelerate or fully automate the processing, quantification and analysis of large amounts of high-information-content biomedical imagery. Modern image analysis systems augment an observer's ability to make measurements from a large or complex set of images, by improving accuracy, objectivity, or speed. A fully developed analysis system may completely replace the observer. Although these systems are not unique to biomedical imagery, biomedical imaging is becoming more important for both diagnostics and research. Some examples are:

- high-throughput and high-fidelity quantification and sub-cellular localization (high-content screening, cytohistopathology, Bioimage informatics)

- morphometrics

- clinical image analysis and visualization

- determining the real-time air-flow patterns in breathing lungs of living animals

- quantifying occlusion size in real-time imagery from the development of and recovery during arterial injury

- making behavioral observations from extended video recordings of laboratory animals

- infrared measurements for metabolic activity determination

- inferring clone overlaps in DNA mapping, e.g. the Sulston score

High-throughput Single Cell Data Analysis

Computational techniques are used to analyse high-throughput, low-measurement single cell data, such as that obtained from flow cytometry. These methods typically involve finding populations of cells that are relevant to a particular disease state or experimental condition.

Biodiversity Informatics

Biodiversity informatics deals with the collection and analysis of biodiversity data, such as taxonomic databases, or microbiome data. Examples of such analyses include phylogenetics, niche modelling, species richness mapping, DNA barcoding, or species identification tools.

Databases

Databases are essential for bioinformatics research and applications. Many databases exist, covering various information types: for example, DNA and protein sequences, molecular structures, phenotypes and biodiversity. Databases may contain empirical data (obtained directly from experiments), predicted data (obtained from analysis), or, most commonly, both. They may be specific to a particular organism, pathway or molecule of interest. Alternatively, they can incorporate data compiled from multiple other databases. These databases vary in their format, access mechanism, and whether they are public or not.

Some of the most commonly used databases are listed below. For a more comprehensive list, please check the link at the beginning of the subsection.

- Used in biological sequence analysis: Genbank, UniProt

- Used in finding Protein Families and Motif Finding: InterPro, Pfam

- Used for Next Generation Sequencing: Sequence Read Archive

- Used in Network Analysis: Metabolic Pathway Databases (KEGG), Interaction Analysis Databases, Functional Networks

- Used in design of synthetic genetic circuits: GenoCAD

Software and Tools

Software tools for bioinformatics range from simple command-line tools, to more complex graphical programs and standalone web-services available from various bioinformatics companies or public institutions.

Open-source Bioinformatics Software

Many free and open-source software tools have existed and continued to grow since the 1980s. The combination of a continued need for new algorithms for the analysis of emerging types of biological readouts, the potential for innovative *in silico* experiments, and freely available open code bases have helped to create opportunities for all research groups to contribute to both bioinformatics and the range of open-source software available, regardless of their funding arrangements. The open source tools often act as incubators of ideas, or community-supported plug-ins in commercial applications. They may also provide *de facto* standards and shared object models for assisting with the challenge of bioinformation integration.

The range of open-source software packages includes titles such as Bioconductor, BioPerl, Biopython, BioJava, BioJS, BioRuby, Bioclipse, EMBOSS,.NET Bio, Orange with its bioinformatics add-on, Apache Taverna, UGENE and GenoCAD. To maintain this tradition and create further opportunities, the non-profit Open Bioinformatics Foundation have supported the annual Bioinformatics Open Source Conference (BOSC) since 2000.

An alternative method to build public bioinformatics databases Additional Reading. This system allows the database to be accessed and updated by all experts in the field.

Web Services in Bioinformatics

SOAP- and REST-based interfaces have been developed for a wide variety of bioinformatics applications allowing an application running on one computer in one part of the world to use algorithms, data and computing resources on servers in other parts of the world. The main advantages derive from the fact that end users do not have to deal with software and database maintenance overheads.

Basic bioinformatics services are classified by the EBI into three categories: SSS (Sequence Search Services), MSA (Multiple Sequence Alignment), and BSA (Biological Sequence Analysis). The availability of these service-oriented bioinformatics resources demonstrate the applicability of web-based bioinformatics solutions, and range from a collection of standalone tools with a common data format under a single, standalone or web-based interface, to integrative, distributed and extensible bioinformatics workflow management systems.

Bioinformatics Workflow Management Systems

A Bioinformatics workflow management system is a specialized form of a workflow management system designed specifically to compose and execute a series of computational or data manipulation steps, or a workflow, in a Bioinformatics application. Such systems are designed to

- provide an easy-to-use environment for individual application scientists themselves to create their own workflows

- provide interactive tools for the scientists enabling them to execute their workflows and view their results in real-time

- simplify the process of sharing and reusing workflows between the scientists.

- enable scientists to track the provenance of the workflow execution results and the workflow creation steps.

Some of the platforms giving this service: Galaxy, Kepler, Taverna, UGENE, Anduril.

Education Platforms

Software platforms designed to teach bioinformatics concepts and methods include Rosalind and online courses offered through the Swiss Institute of Bioinformatics Training Portal. The Canadian Bioinformatics Workshops provides videos and slides from training workshops on their website under a Creative Commons license. The 4273π project or 4273pi project also offers open source educational materials for free. The course runs on low cost raspberry pi computers and has been used to teach adults and school pupils. 4273π is actively developed by a consortium of academics and research staff who have run research level bioinformatics using raspberry pi computers and the 4273π operating system.

MOOC platforms also provide online certifications in bioinformatics and related disciplines, including Coursera's Bioinformatics Specialization (UC San Diego) and Genomic Data Science Specialization (Johns Hopkins) as well as EdX's Data Analysis for Life Sciences XSeries (Harvard).

Conferences

There are several large conferences that are concerned with bioinformatics. Some of the most notable examples are Intelligent Systems for Molecular Biology (ISMB), European Conference on Computational Biology (ECCB), and Research in Computational Molecular Biology (RECOMB).

Kyoto Encyclopedia of Genes and Genomes

KEGG (Kyoto Encyclopedia of Genes and Genomes) is a collection of databases dealing with genomes, biological pathways, diseases, drugs, and chemical substances. KEGG is utilized for bioinformatics research and education, including data analysis in genomics, metagenomics, metabolomics and other omics studies, modeling and simulation in systems biology, and translational research in drug development.

Introduction

The KEGG database project was initiated in 1995 by Minoru Kanehisa, Professor at the Institute for Chemical Research, Kyoto University, under the then ongoing Japanese Human Genome Program. Foreseeing the need for a computerized resource that can be used for biological interpretation of genome sequence data, he started developing the KEGG PATHWAY database. It is a collection of manually drawn KEGG pathway maps representing experimental knowledge on metabolism and various other functions of the cell and the organism. Each pathway map contains a network of molecular interactions and reactions and is designed to link genes in the genome to gene products (mostly proteins) in the pathway. This has enabled the analysis called KEGG pathway mapping, whereby the gene content in the genome is compared with the KEGG PATHWAY database to examine which pathways and associated functions are likely to be encoded in the genome.

According to the developers, KEGG is a "computer representation" of the biological system. It integrates building blocks and wiring diagrams of the system — more specifically, genetic building blocks of genes and proteins, chemical building blocks of small molecules and reactions, and wiring diagrams of molecular interaction and reaction networks. This concept is realized in the following databases of KEGG, which are categorized into systems, genomic, chemical, and health information.

- Systems information
    - PATHWAY — pathway maps for cellular and organismal functions
    - MODULE — modules or functional units of genes
    - BRITE — hierarchical classifications of biological entities
- Genomic information
    - GENOME — complete genomes
    - GENES — genes and proteins in the complete genomes
    - ORTHOLOGY — ortholog groups of genes in the complete genomes
- Chemical information
    - COMPOUND, GLYCAN — chemical compounds and glycans
    - REACTION, RPAIR, RCLASS — chemical reactions
    - ENZYME — enzyme nomenclature
- Health information
    - DISEASE — human diseases

      o   DRUG — approved drugs

      o   ENVIRON — crude drugs and health-related substances

Databases

Systems Information

The KEGG PATHWAY database, the wiring diagram database, is the core of the KEGG resource. It is a collection of pathway maps integrating many entities including genes, proteins, RNAs, chemical compounds, glycans, and chemical reactions, as well as disease genes and drug targets, which are stored as individual entries in the other databases of KEGG. The pathway maps are classified into the following sections:

- Metabolism

- Genetic information processing (transcription, translation, replication and repair, etc.)

- Environmental information processing (membrane transport, signal transduction, etc.)

- Cellular processes (cell growth, cell death, cell membrane functions, etc.)

- Organismal systems (immune system, endocrine system, nervous system, etc.)

- Human diseases

- Drug development

The metabolism section contains aesthetically drawn global maps showing an overall picture of metabolism, in addition to regular metabolic pathway maps. The low-resolution global maps can be used, for example, to compare metabolic capacities of different organisms in genomics studies and different environmental samples in metagenomics studies. In contrast, KEGG modules in the KEGG MODULE database are higher-resolution, localized wiring diagrams, representing tighter functional units within a pathway map, such as subpathways conserved among specific organism groups and molecular complexes. KEGG modules are defined as characteristic gene sets that can be linked to specific metabolic capacities and other phenotypic features, so that they can be used for automatic interpretation of genome and metagenome data.

Another database that supplements KEGG PATHWAY is the KEGG BRITE database. It is an ontology database containing hierarchical classifications of various entities including genes, proteins, organisms, diseases, drugs, and chemical compounds. While KEGG PATHWAY is limited to molecular interactions and reactions of these entities, KEGG BRITE incorporates many different types of relationships.

Genomic Information

Several months after the KEGG project was initiated in 1995, the first report of the completely sequenced bacterial genome was published. Since then all published complete genomes are accumulated in KEGG for both eukaryotes and prokaryotes. The KEGG GENES database contains gene/protein-level information and the KEGG GENOME database contains organism-level information for these genomes. The KEGG GENES database consists of gene sets for the complete genomes, and genes in each set are given annotations in the form of establishing correspondences to the wiring diagrams of KEGG pathway maps, KEGG modules, and BRITE hierarchies.

These correspondences are made using the concept of orthologs. The KEGG pathway maps are drawn based on experimental evidence in specific organisms but they are designed to be applicable to other organisms as well, because different organisms, such as human and mouse, often share identical pathways consisting of functionally identical genes, called orthologous genes or orthologs. All the genes in the KEGG GENES database are being grouped into such orthologs in the KEGG ORTHOLOGY (KO) database. Because the nodes (gene products) of KEGG pathway maps, as well as KEGG modules and BRITE hierarchies, are given KO identifiers, the correspondences are established once genes in the genome are annotated with KO identifiers by the genome annotation procedure in KEGG.

Chemical Information

The KEGG metabolic pathway maps are drawn to represent the dual aspects of the metabolic network: the genomic network of how genome-encoded enzymes are connected to catalyze consecutive reactions and the chemical network of how chemical structures of substrates and products are transformed by these reactions. A set of enzyme genes in the genome will identify enzyme relation networks when superimposed on the KEGG pathway maps, which in turn characterize chemical structure transformation networks allowing interpretation of biosynthetic and biodegradation potentials of the organism. Alternatively, a set of metabolites identified in the metabolome will lead to the understanding of enzymatic pathways and enzyme genes involved.

The databases in the chemical information category, which are collectively called KEGG LIGAND, are organized by capturing knowledge of the chemical network. In the beginning of the KEGG project, KEGG LIGAND consisted of three databases: KEGG COMPOUND for chemical compounds, KEGG REACTION for chemical reactions, and KEGG ENZYME for reactions in the enzyme nomenclature. Currently, there are additional databases: KEGG GLYCAN for glycans and two auxiliary reaction databases called RPAIR (reactant pair alignments) and RCLASS (reaction class). KEGG COMPOUND has also been expanded to contain various compounds such as xenobiotics, in addition to metabolites.

Health Information

In KEGG, diseases are viewed as perturbed states of the biological system caused by perturbants of genetic factors and environmental factors, and drugs are viewed as different types of perturbants. The KEGG PATHWAY database includes not only the normal states but also the perturbed states of the biological systems. However, disease pathway maps cannot be drawn for most diseases because molecular mechanisms are not well understood. An alternative approach is taken in the KEGG DISEASE database, which simply catalogs known genetic factors and environmental factors of diseases. These catalogs may eventually lead to more complete wiring diagrams of diseases.

The KEGG DRUG database contains active ingredients of approved drugs in Japan, the USA, and Europe. They are distinguished by chemical structures and/or chemical components and associated with target molecules, metabolizing enzymes, and other molecular interaction network information in the KEGG pathway maps and the BRITE hierarchies. This enables an integrated analysis of drug interactions with genomic information. Crude drugs and other health-related substances, which are outside of the category of approved drugs, are stored in the KEGG ENVIRON database. The databases in the health information category are collectively called KEGG MEDICUS, which also includes package inserts of all marketed drugs in Japan.

Subscription Model

In July 2011 KEGG introduced a subscription model for FTP download due to a significant cutback of government funding. KEGG continues to be freely available through its website, but the subscription model has raised discussions about sustainability of bioinformatics databases.

Kegg Ligand

KEGG LIGAND comprises four databases: COMPOUND, GLYCAN, REACTION, and ENZYME. COMPOUND is a database of chemical structures of most known metabolic compounds and some pharmaceutical and environmental compounds; GLYCAN is a database of carbohydrate structures; REACTION is a database of reaction formulas for enzymic reactions; and ENZYME is a database of enzyme nomenclatures.

Kegg Brite

KEGG BRITE gives the functional hierarchies representing our knowledge on various aspects of biological systems.

Cytoscape

Cytoscape is an open source bioinformatics software platform for visualizing molecular interaction networks and integrating with gene expression profiles and other state data. Additional features are available as plugins. Plugins are available for network and molecular profiling analyses, new layouts, additional file format support and connection with databases and searching in large networks. Plugins may be developed using the Cytoscape open Java software architecture by anyone and plugin community development is encouraged. Cytoscape also has a JavaScript-centric sister project named Cytoscape.js that can be used to analyse and visualise graphs in JavaScript environments, like a browser.

History

Cytoscape was originally created at the Institute of Systems Biology in Seattle in 2002. Now, it is developed by an international consortium of open source developers. Cytoscape was initially made public in July, 2002 (v0.8); the second release (v0.9) was in November, 2002, and v1.0 was released in March 2003. Version 1.1.1 is the last stable release for the 1.0 series. Version 2.0 was initially released in 2004; Cytoscape 2.83, the final 2.xx version, was released in May 2012. Version 3.0 was released Feb 1, 2013, and the latest version, 3.4.0, was released in May 2016.

Development

The Cytoscape core developer team continues to work on this project and released Cytoscape 3.0 in 2013. This represented a major change in the Cytoscape architecture; it is a more modularized, expandable and maintainable version of the software. As of February 2015, work is beginning on version 3.3.

Usage

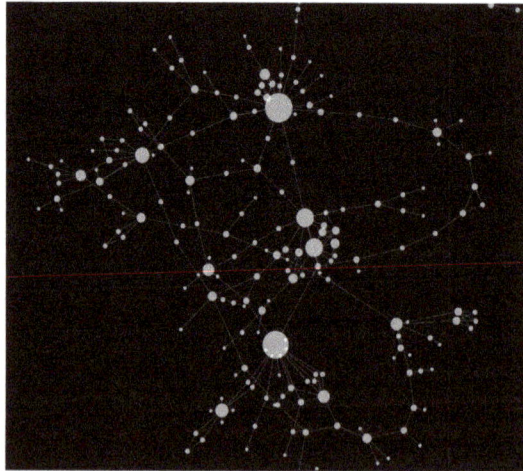

Yeast Protein–protein/Protein–DNA interaction network visualized by Cytoscape.
Node degree is mapped to node size

While Cytoscape is most commonly used for biological research applications, it is agnostic in terms of usage. Cytoscape can be used to visualize and analyze network graphs of any kind involving nodes and edges (e.g., social networks). A key aspect of the software architecture of Cytoscape is the use of plugins for specialized features. Plugins are developed by core developers and the greater user community.

Features

Input

- Input and construct molecular interaction networks from raw interaction files (SIF format) containing lists of protein–protein and/or protein–DNA interaction pairs. For yeast and other model organisms, large sources of pairwise interactions are available through the BIND and TRANSFAC databases. User-defined interaction types are also supported.

- Load and save previously-constructed interaction networks in GML format (Graph Modelling Language).

- Load and save networks and node/edge attributes in an XML document format called XGMML (eXtensible Graph Markup and Modeling Language).

- Input mRNA expression profiles from tab- or space-delimited text files.

- Load and save arbitrary attributes on nodes and edges. For example, input a set of custom annotation terms for your proteins, create a set of confidence values for your protein–protein interactions.

- Import gene functional annotations from the Gene Ontology (GO) and KEGG databases.

- Directly import GO terms and annotations from OBO and Gene Association files.

- Load and save state of the cytoscape session in a cytoscape session (.cys) file. Cytoscape session file includes networks, attributes (for node/edge/network), desktop states (selected/hidden nodes and edges, window sizes), properties, and visual styles.

Visualization

- Customize network data display using powerful visual styles.

- View a superposition of gene expression ratios and p-values on the network. Expression data can be mapped to node color, label, border thickness, or border color, etc. according to user-configurable colors and visualization schemes.

- Layout networks in two dimensions. A variety of layout algorithms are available, including cyclic and spring-embedded layouts.

- Zoom in/out and pan for browsing the network.

- Use the network manager to easily organize multiple networks. And this structure can be saved in a session file.

- Use the bird's eye view to easily navigate large networks.

- Easily navigate large networks (100,000+ nodes and edges) by efficient rendering engine.

Analysis

- Plugins available for network and molecular profile analysis. For example:

    o   Filter the network to select subsets of nodes and/or interactions based on the current data. For instance, users may select nodes involved in a threshold number of interactions, nodes that share a particular GO annotation, or nodes whose gene expression levels change significantly in one or more conditions according to p-values loaded with the gene expression data.

    o   Find active subnetworks/pathway modules. The network is screened against gene expression data to identify connected sets of interactions, i.e. interaction subnetworks, whose genes show particularly high levels of differential expression. The interactions contained in each subnetwork provide hypotheses for the regulatory and signaling interactions in control of the observed expression changes.

    o   Find clusters (highly interconnected regions) in any network loaded into Cytoscape. Depending on the type of network, clusters may mean different things. For instance, clusters in a protein–protein interaction network have been shown to be protein complexes and parts of pathways. Clusters in a protein similarity network represent protein families.

One can choose from different formats to input the molecular and genetic interaction data sets to the tool. The tool also helps in visualisation and analysis of human-curated pathway datasets from KEGG.

Cytoscape is a Java application which can run on Linux, Windows, and Mac OS X.

Cytoscape modules allow us to create new networks. A new network can be created from an old network, through the option (via the File → New → Network option),

The new network will be indicated as the child of the parent network form which it was derived.

The Web Service Client Manager module of Cytoscape helps users in accessing various kinds of databases.

Networks can be created in Cytoscape using any of the following methods:

1. Importing pre-existing, formatted network files.

2. Importing pre-existing, unformatted text or Excel files.

3. Importing networks from Web Service.

4. Creating an empty network and manually adding nodes and edges.

Cytoscape is aligned to read network/pathway files written in any of the following formats:

- Simple interaction file (SIF or.sif format)

- Nested network format (NNF or.nnf format)

- Graph Markup Language (GML or.gml format)

- XGMML (extensible graph markup and modelling language).

- SBML

- BioPAX

- PSI-MI Level 1 and 2.5

- Delimited text

- Excel Workbook (.xls)

Creating Network using Cytoscape

1. Go to file->New->network->empty network.

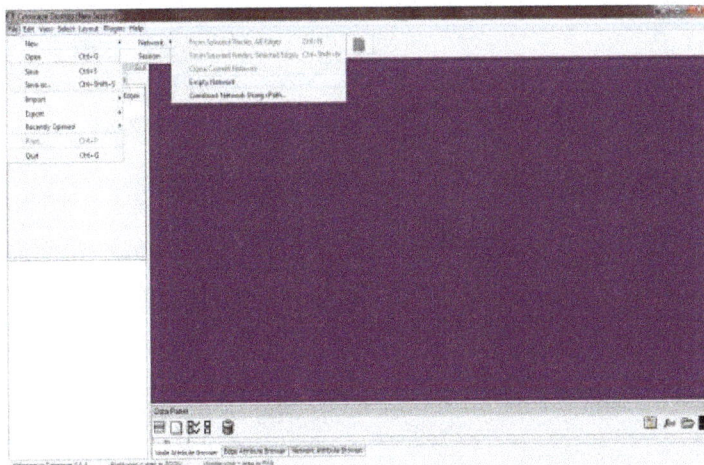

2. A new network window is opened now.

3. In the control panel under editor the new nodes , edges can be created by following ways,

Drag and Drop:

- A node shape onto the network view.

- An edge shape onto the source node, then click on the target node.

Double-click:

- To add nodes and edges specified in SIF format

CTRL-click:

- On empty space to create a node.

- On a node to begin an edge and specify the source node. Then click on the target node to finish the edge.

4. In the control panel the name of the network can be edited and changed. We can specify the number of nodes and edges or else we can add the node manually.

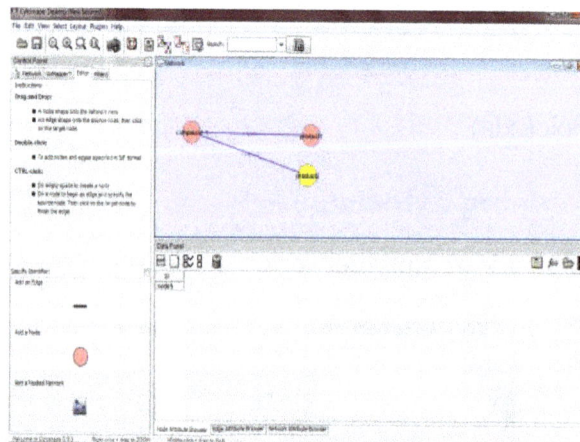

5. The network can be viewed by exporting into the graphical format as,

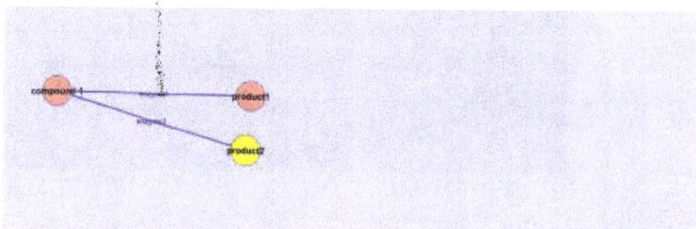

6. We can also change the color and the size of the nodes by right clicking on the node and changing the node color and the node size and the label can also be added accordingly.

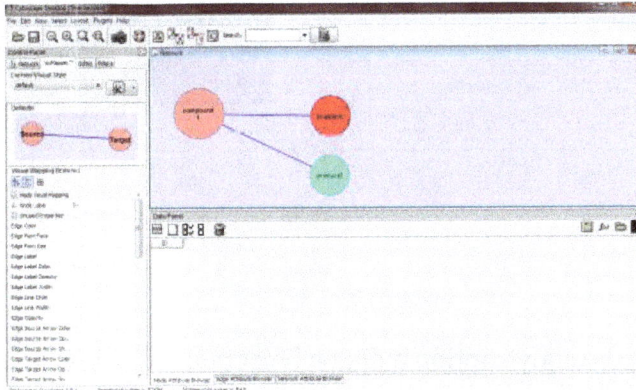

7. Thus the more complex network can be remodelled manually by Cytoscape.

8. Consider a part of the KEGG pathway we created (eg. glucose) and find the interaction by creating a network. Figure gives a part of the network to show how the network is made.

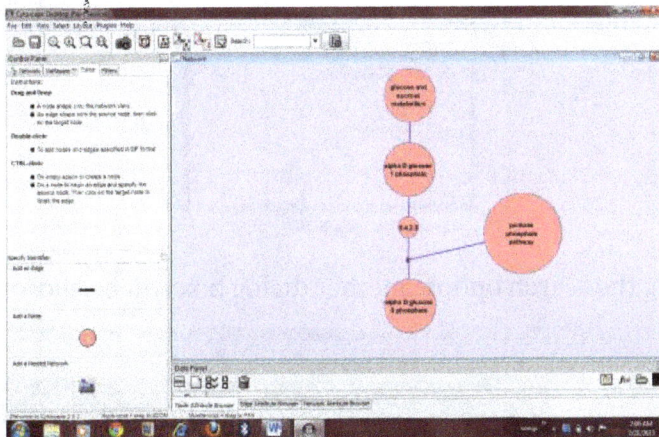

Interaction network of a KEGG pathway

➤ Draw a pathway map using Cytoscape for the gene p53.

We shall now detail import of files and development of new networks.

Mapping Networks

➤ The network can be imported from ncbi and several other databases using cytoscape.

➤ This can be done by File => Import => Network from web services.

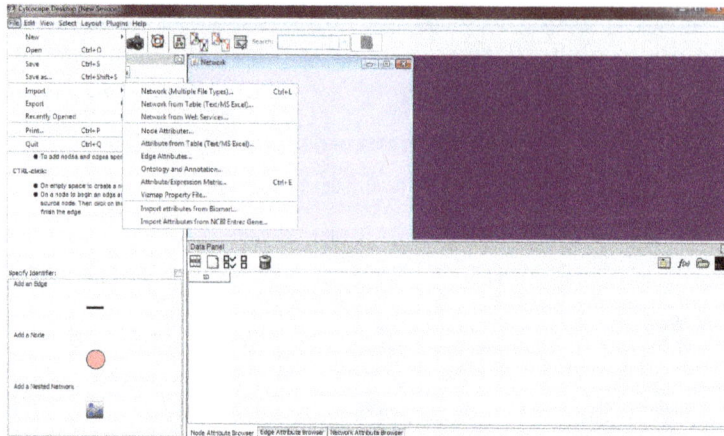

> By clicking the above option a dialog box opens as below. In this dialog box the gene name of the organism can be searched.

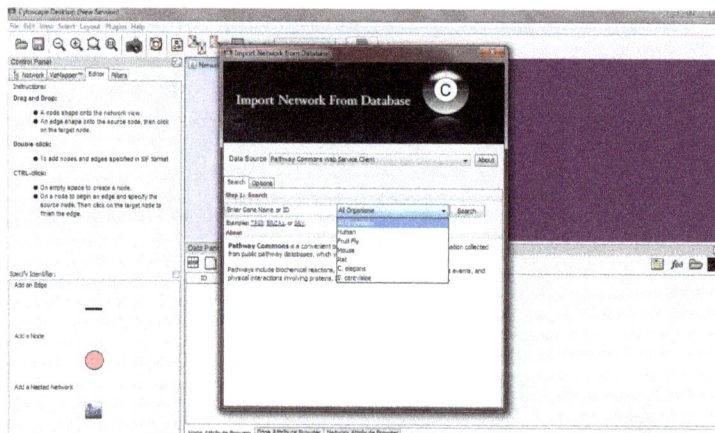

> By clicking the search option another dialog box will be shown as below.

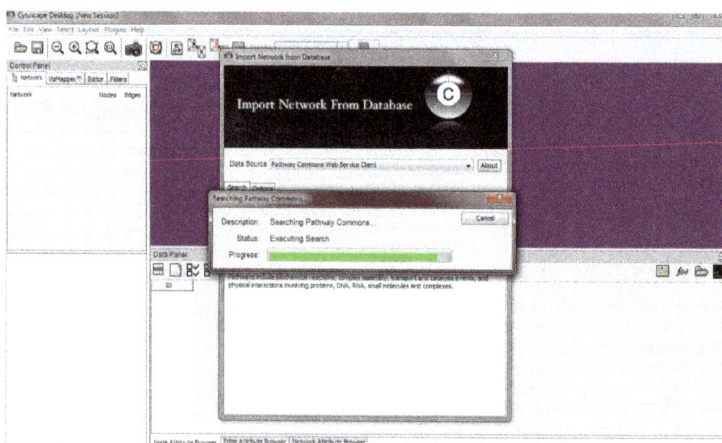

> In several pathways listed the relevant pathway can be selected.

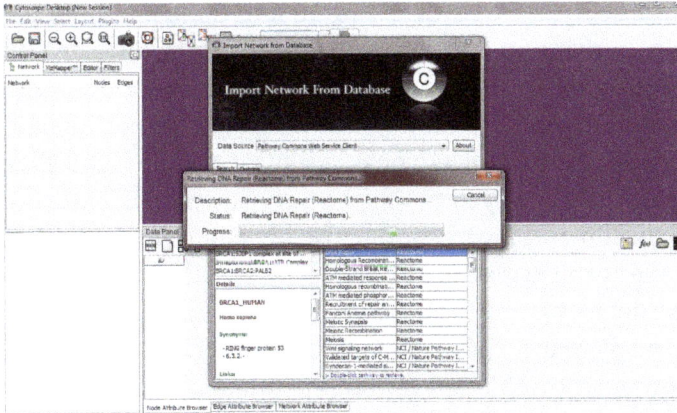

> ➤ After selecting the pathway the page takes and imports the network as shown below.

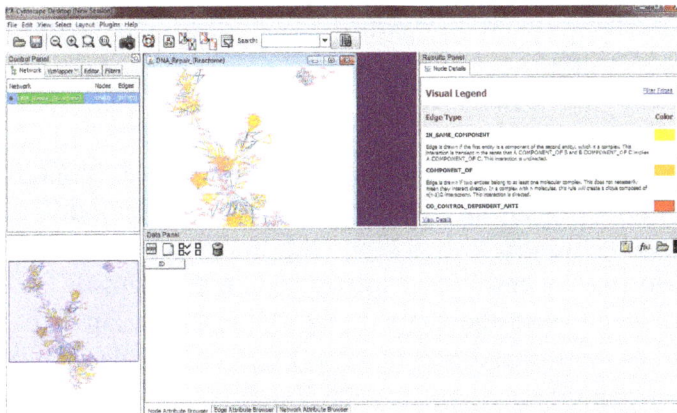

> ➤ By using the above steps the pathway for apoptopic gene and p53 gene can be retrieved. Here we can modify the network that is already available and can edit the interactions.

SBML

The Systems Biology Markup Language (SBML) is a representation format, based on XML, for communicating and storing computational models of biological processes. It is a free and open standard with widespread software support and a community of users and developers. SBML can represent many different classes of biological phenomena, including metabolic networks, cell signaling pathways, regulatory networks, infectious diseases, and many others. It has been proposed as a standard for representing computational models in systems biology today.

History

Late in the year 1999 through early 2000, with funding from the Japan Science and

Technology Corporation (JST), Hiroaki Kitano and John C. Doyle assembled a small team of researchers to work on developing better software infrastructure for computational modeling in systems biology. Hamid Bolouri was the leader of the development team, which consisted of Andrew Finney, Herbert Sauro, and Michael Hucka. Bolouri identified the need for a framework to enable interoperability and sharing between the different simulation software systems for biology in existence during the late 1990s, and he organized an informal workshop in December 1999 at the California Institute of Technology to discuss the matter. In attendance at that workshop were the groups responsible for the development of DBSolve, E-Cell, Gepasi, Jarnac, StochSim and The Virtual Cell. Separately, earlier in 1999, some members of these groups also had discussed the creation of a portable file format for metabolic network models in the Bio-ThermoKinetics (BTK) group. The same groups who attended the first Caltech workshop met again on April 28–29, 2000, at the first of a newly created meeting series called *Workshop on Software Platforms for Systems Biology*. It became clear during the second workshop that a *common model representation format* was needed to enable the exchange of models between software tools as part of any functioning interoperability framework, and the workshop attendees decided the format should be encoded in XML.

The Caltech ERATO team developed a proposal for this XML-based format and circulated the draft definition to the attendees of the 2nd Workshop on Software Platforms for Systems Biology in August, 2000. This draft underwent extensive discussion over mailing lists and during the 2nd Workshop on Software Platforms for Systems Biology, held in Tokyo, Japan, in November 2000 as a satellite workshop of the ICSB 2000 conference. After further revisions, discussions and software implementations, the Caltech team issued a specification for SBML Level 1, Version 1 in March, 2001.

SBML Level 2 was conceived at the 5th Workshop on Software Platforms for Systems Biology, held in July 2002, at the University of Hertfordshire, UK. By this time, far more people were involved than the original group of SBML collaborators and the continued evolution of SBML became a larger community effort, with many new tools having been enhanced to support SBML. The workshop participants in 2002 collectively decided to revise the form of SBML in Level 2. The first draft of the Level 2 Version 1 specification was released in August 2002, and the final set of features was finalized in May 2003 at the 7th Workshop on Software Platforms for Systems Biology in Ft. Lauderdale, Florida.

The next iteration of SBML took two years in part because software developers requested time to absorb and understand the larger and more complex SBML Level 2. The inevitable discovery of limitations and errors led to the development of SBML Level 2 Version 2, issued in September 2006. By this time, the team of SBML Editors (who reconcile proposals for changes and write a coherent final specification document) had changed and now consisted of Andrew Finney, Michael Hucka and Nicolas Le Novère.

SBML Level 2 Version 3 was published in 2007 after countless contributions by and discussions with the SBML community. 2007 also saw the election of two more SBML Editors as part of the introduction of the modern SBML Editor organization in the context of the SBML development process.

SBML Level 2 Version 4 was published in 2008 after certain changes in Level 2 were requested by popular demand. (For example, an electronic vote by the SBML community in late 2007 indicated a majority preferred not to require strict unit consistency before an SBML model is considered valid.) Version 4 was finalized after the SBML Forum meeting held in Gothenburg, Sweden, as a satellite workshop of ICSB 2008 in the fall of 2008.

SBML Level 3 Version 1 Core was published in final form in 2010, after prolonged discussion and revision by the SBML Editors and the SBML community. It contains numerous significant changes in syntax and constructs from Level 2 Version 4, but also represents a new modular base for continued expansion of SBML's features and capabilities going into the future.

SBML Level 2 Version 5 was published in 2015. This revision included a number of textual (but not structural) changes in response to user feedback, thereby addressing the list of errata collected over many years for the SBML Level 2 Version 4 specification. In addition, Version 5 introduced a facility to use nested annotations within SBML's annotation format (an annotation format that is based on a subset of RDF).

The Language

SBML is sometimes incorrectly assumed to be limited in scope only to biochemical network models because the original publications and early software focused on this domain. In reality, although the central features of SBML are indeed oriented towards representing chemical reaction-like processes that act on entities, this same formalism serves analogously for many other types of processes; moreover, SBML has language features supporting the direct expression of mathematical formulas and discontinuous events separate from reaction processes, allowing SBML to represent much more than solely biochemical reactions. Evidence for SBML's ability to be used for more than merely descriptions of biochemistry can be seen in the variety of models available from BioModels Database.

Purposes

SBML has three main purposes:

- enable the use of multiple software tools without having to rewrite models to conform to every tool's idiosyncratic file format;

- enable models to be shared and published in a form that other researchers can use even when working with different software environments;

- ensure the survival of models beyond the lifetime of the software used to create them.

SBML is not an attempt to define a universal language for quantitative models. SBML's purpose is to serve as a *lingua franca*—an exchange format used by different present-day software tools to communicate the essential aspects of a computational model.

Main Capabilities

SBML can encode models consisting of entities (called *species* in SBML) acted upon by processes (called *reactions*). An important principle is that models are decomposed into explicitly-labeled constituent elements, the set of which resembles a verbose rendition of chemical reaction equations (if the model uses reactions) together with optional explicit equations (again, if the model uses these); the SBML representation deliberately does not cast the model directly into a set of differential equations or other specific interpretation of the model. This explicit, modeling-framework-agnostic decomposition makes it easier for a software tool to interpret the model and translate the SBML form into whatever internal form the tool actually uses.

A software package can read an SBML model description and translate it into its own internal format for model analysis. For example, a package might provide the ability to simulate the model by constructing differential equations and then perform numerical time integration on the equations to explore the model's dynamic behavior. Or, alternatively, a package might construct a discrete stochastic representation of the model and use a Monte Carlo simulation method such as the Gillespie algorithm.

SBML allows models of arbitrary complexity to be represented. Each type of component in a model is described using a specific type of data structure that organizes the relevant information. The data structures determine how the resulting model is encoded in XML.

In addition to the elements above, another important feature of SBML is that every entity can have machine-readable annotations attached to it. These annotations can be used to express relationships between the entities in a given model and entities in external resources such as databases. A good example of the value of this is in BioModels Database, where every model is annotated and linked to relevant data resources such as publications, databases of compounds and pathways, controlled vocabularies, and more. With annotations, a model becomes more than simply a rendition of a mathematical construct—it becomes a semantically-enriched framework for communicating knowledge.

Levels and Versions

SBML is defined in Levels: upward-compatible specifications that add features and expressive power. Software tools that do not need or cannot support the complexity of higher Levels can go on using lower Levels; tools that can read higher Levels are assured of also being able to interpret models defined in the lower Levels. Thus new Levels do not supersede previous ones. However, each Level can have multiple Versions within it, and new Versions of a Level do supersede old Versions of that same Level.

There are currently three Levels of SBML defined. The current Versions within those Levels are the following:

- Level 3 Version 1 Core, for which the final Release 1 specification was issued 6 October 2010

- Level 2 Version 5 Release 1

- Level 1 Version 2

Open-source software infrastructure such as libSBML and JSBML allows developers to support all Levels of SBML their software with a minimum amount of effort.

The SBML Team maintains a public issue tracker where readers may report errors or other issues in the SBML specification documents. Reported issues are eventually put on the list of official errata associated with each specification release. The lists of errata are documented on the Specifications page of SBML.org.

Level 3 Packages

Development of SBML Level 3 has been proceeding in a modular fashion. The *Core* specification is a complete format that can be used alone. Additional Level 3 packages can be layered on to this core to provide additional, optional features.

Hierarchical Model Composition

The Hierarchical Model Composition package, known as "*comp*", was released in November 2012. This package provides the ability to include models as submodels inside another model. The goal is to support the ability of modelers and software tools to do such things as (1) decompose larger models into smaller ones, as a way to manage complexity; (2) incorporate multiple instances of a given model within one or more enclosing models, to avoid literal duplication of repeated elements; and (3) create libraries of reusable, tested models, much as is done in software development and other engineering fields. The specification was the culmination of years of discussion by a wide number of people.

Flux Balance Constraints

The Flux Balance Constraints package (nicknamed "*fbc*") was first released in February, 2013. Import revisions were introduced as part of Version 2, released in September, 2015. The "*fbc*" package provides support for constraint based modeling, frequently used to analyze and study biological networks on both a small and large scale. This SBML package makes use of standard components from the SBML Level 3 core specification, including species and reactions, and extends them with additional attributes and structures to allow modelers to define such things as flux bounds and optimization functions.

Qualitative Models

The Qualitative Models or "*qual*" package for SBML Level 3 was released in May, 2013. This package supports the representation of models where an in-depth knowledge of the biochemical reactions and their kinetics is missing and a qualitative approach must be used. Examples of phenomena that have been modeled in this way include gene regulatory networks and signalling pathways, basing the model structure on the definition of regulatory or influence graphs. The definition and use of some components of this class of models differ from the way that species and reactions are defined and used in *core* SBML models. For example, qualitative models typically associate discrete levels of activities with entity pools; consequently, the processes involving them cannot be described as reactions per se, but rather as transitions between states. These systems can be viewed as reactive systems whose dynamics are represented by means of state transition graphs (or other Kripke structures) in which the nodes are the reachable states and the edges are the state transitions.

Layout

The SBML *layout* package originated as a set of annotation conventions usable in SBML Level 2. It was introduced at the SBML Forum in St. Louis in 2004. Ralph Gauges wrote the specification and provided an implementation that was widely used. This original definition was reformulated as an SBML Level 3 package, and a specification was formally released in August, 2013.

The SBML Level 3 Layout package provides a specification for how to represent a reaction network in a graphical form. It is thus better tailored to the task than the use of an arbitrary drawing or graph. The SBML Level 3 package only deals with the information necessary to define the position and other aspects of a graph's layout; the additional details necessary to complete the graph—namely, how the visual aspects are meant to be rendered— are the purvey of the separate SBML Level 3 package called *Rendering* (nicknamed "*render*"). As of November 2015, a draft specification for the "*render*" package is available, but it has not yet been officially finalized.

Packages Under Development

Development of SBML Level 3 packages is being undertaken such that specifications are reviewed and implementations attempted during the development process. Once a specification is a stable and there are two implementations that support it, the package is considered accepted. The packages detailed above have all reached the accepted stage. The table below gives a brief summary of packages that are currently in the development phase.

Package name	Label	Description
Arrays	arrays	Support for expressing arrays of components
Distributions	distrib	Support for encoding models that sample values from statistical distributions or specify statistics associated with numerical values
Dynamics	dyn	Support for creating and destroying entities during a simulation
Groups	groups	A means for grouping elements
Multistate and Multi-component species	multi	Object structures for representing entity pools with multiple states and composed of multiple components, and reaction rules involving them
Rendering	render	Support for defining the graphical symbols and glyphs used in a diagram of the model; adjunct to the layout package
Required Elements	req	Support for fine-grained indication of SBML elements that have been changed by the presence of another package
Spatial Processes	spatial	Support for describing processes that involve a spatial component, and describing the geometries involved

Structure

A model definition in SBML Levels 2 and 3 consists of lists of one or more of the following components:

- Function definition: A named mathematical function that may be used throughout the rest of a model.

- Unit definition: A named definition of a new unit of measure, or a redefinition of an existing SBML default unit. Named units can be used in the expression of quantities in a model.

- Compartment Type (only in SBML Level 2): A type of location where reacting entities such as chemical substances may be located.

- Species type (only in SBML Level 2): A type of entity that can participate in reactions. Examples of species types include ions such as Ca^{2+}, molecules such as glucose or ATP, binding sites on a protein, and more.

- Compartment: A well-stirred container of a particular type and finite size where species may be located. A model may contain multiple compartments of the same compartment type. Every species in a model must be located in a compartment.

- Species: A pool of entities of the same species type located in a specific compartment.

- Parameter: A quantity with a symbolic name. In SBML, the term parameter is used in a generic sense to refer to named quantities regardless of whether they are constants or variables in a model.

- Initial Assignment: A mathematical expression used to determine the initial conditions of a model. This type of structure can only be used to define how the value of a variable can be calculated from other values and variables at the start of simulated time.

- Rule: A mathematical expression used in combination with the differential equations constructed based on the set of reactions in a model. It can be used to define how a variable's value can be calculated from other variables, or used to define the rate of change of a variable. The set of rules in a model can be used with the reaction rate equations to determine the behavior of the model with respect to time. The set of rules constrains the model for the entire duration of simulated time.

- Constraint: A mathematical expression that defines a constraint on the values of model variables. The constraint applies at all instants of simulated time. The set of constraints in model should not be used to determine the behavior of the model with respect to time.

- Reaction: A statement describing some transformation, transport or binding process that can change the amount of one or more species. For example, a reaction may describe how certain entities (reactants) are transformed into certain other entities (products). Reactions have associated kinetic rate expressions describing how quickly they take place.

- Event: A statement describing an instantaneous, discontinuous change in a set of variables of any type (species concentration, compartment size or parameter value) when a triggering condition is satisfied.

Community

As of September, 2015, more than 280 software systems advertise support for SBML. A current list is available in the form of the SBML Software Guide, hosted at SBML.org.

SBML has been and continues to be developed by the community of people making software platforms for systems biology, through active email discussion lists and biannual workshops. The meetings are often held in conjunction with other biology conferences, especially the International Conference on Systems Biology (ICSB). The community effort is coordinated by an elected editorial board made up of five members. Each editor is elected for a 3-year non-renewable term.

Tools such as an online model validator as well as open-source libraries for incorporating SBML into software programmed in the C, C++, Java, Python, Mathematica,

MATLAB and other languages are developed partly by the SBML Team and partly by the broader SBML community.

SBML is an official IETF MIME type, specified by RFC 3823.

TRANSFAC

TRANSFAC (TRANScription FACtor database) is a manually curated database of eukaryotic transcription factors, their genomic binding sites and DNA binding profiles. The contents of the database can be used to predict potential transcription factor binding sites.

Introduction

The origin of the database was an early data collection published 1988. The first version that was released under the name TRANSFAC was developed at the former German National Research Centre for Biotechnology and designed for local installation (now: Helmholtz Centre for Infection Research). In one of the first publicly funded bioinformatics projects, launched in 1993, TRANSFAC developed into a resource that became available on the Internet.

In 1997, TRANSFAC was transferred to a newly established company, BIOBASE, in order to secure long-term financing of the database. Since then, the most up-to-date version has to be licensed, whereas older versions are free for non-commercial users. Since July 2016, TRANSFAC is maintained and distributed by geneXplain GmbH, Wolfenbüttel, Germany.

Content and Features

The content of the database is organized in a way that it is centered around the interaction between transcription factors (TFs) and their DNA binding sites (TFBS). TFs are described with regard to their structural and functional features, extracted from the original scientific literature. They are classified to families, classes and superclasses according to the features of their DNA binding domains.

Binding of a TF to a genomic site is documented by specifying the localization of the site, its sequence and the experimental method applied. All sites that refer to one TF, or a group of closely related TFs, are aligned and used to construct a position-specific scoring matrix (PSSM), or count matrix. Many matrices of the TRANSFAC matrix library have been constructed by a team of curators, others were taken from scientific publications.

Availability

The usage of an older version of TRANSFAC is free of charge for non-profit users. Access to the most up-to-date version requires a license.

Applications

The TRANSFAC database can be used as an encyclopedia of eukaryotic transcription factors. The target sequences and the regulated genes can be listed for each TF, which can be used as benchmark for TFBS recognition tools or as training sets for new TFBS recognition algorithms. The TF classification enables to analyze such data sets with regard to the properties of the DNA-binding domains. Another application is to retrieve all TFs that regulate a given (set of) gene(s). In the context of systems-biological studies, the TF-target gene relations documented in TRANSFAC were used to construct and analyze transcription regulatory networks. By far the most frequent use of TRANSFAC is the computational prediction of potential transcription factor binding sites (TFBS). A number of algorithms exist which either use the individual binding sites or the matrix library for this purpose:

- Patch – analyzes sequence similarities with the binding sites documented in TRANSFAC; it is provided along with the database.

- SiteSeer – analyzes sequence similarities with the binding sites documented in TRANSFAC.

- Match – identifies potential TFBS using the matrix library; it is provided along with the database.

- TESS (Transcription Element Search System) – analyzes sequence similarities with binding sites of TRANSFAC as well as potential binding sites using the matrix libraries of TRANSFAC and three other sources. TESS also provides a program for the identification of cis-regulatory modules (CRMs, characteristic combinations of TFBSs), which uses TRANSFAC matrices.

- PROMO – matrix-based prediction of TFBSs with aid of the commercial database version

- TFM Explorer – Identification of common potential TFBSs in a set of genes

- MotifMogul – matrix-based sequence analysis with a number of different algorithms

- ConTra – matrix-based sequence analysis in conserved promoter regions

- PMS (Poly Matrix Search) – matrix-based sequence analysis in conserved promoter regions

Comparison of matrices with the matrix library of TRANSFAC and other sources:

- T-Reg Comparator to compare individual or groups of matrices with those of TRANSFAC or other libraries.

- MACO (Poly Matrix Search) – matrix comparison with matrix libraries.

A number of servers provide genomic annotations computed with the aid of TRANS-FAC. Others have used such analyses to infer target gene sets.

Similar Data Sources

The following resources offer contents that are related to or partially overlapping with TRANSFAC:

- PlantRegMap - Plant Transcription Factors, cis-elements, binding motifs, and analysis servers

- JASPAR – collection of transcription factor binding profiles (matrices) and sequence analysis program

- PLACE – cis-regulatory DNA elements in plants; until February 2007

- PlantCARE – cis-regulatory elements and transcription factors in plants (2002)

- PRODORIC – a similar concept as TRANSFAC for prokaryotes

- RegTransBase - transcription factor binding sites in a diverse set of bacteria.

- RegulonDB – focus on the bacterium *Escherichia coli*

- SCPD – specific collection of data- and tools for yeast (*Saccharomyces cerevisiae*) (1998)

- TFe – the transcription factor encyclopedia

- TRRD – Transcription Regulatory Regions Database, mainly about regulatory regions and TF-binding sites

- PAZAR - Database with focus on experimentally validated transcription factor binding sites

- Plant Transcription Factor Database and Transcriptional Regulation Data and Analysis Platform

- HOCOMOCO - Homo Sapiens Comprehensive Model Collection

Virtual Cell

This is a beginner's guide to biological model development using VCell/ Virtual Cell. Virtual Cell is a cellular modelling, simulation and analysis software environment developed at Center for Cell Analysis and Modelling (CCAM), University of Connecticut Health Center. It is widely utilized in quantitative cell biological research. It can be utilized to construct mathematical models that range from simple to highly complex experimental data or solely hypothesis driven theoretical modelling of cell biological processes.

It is a web based open-source tool with a single Java based graphical interface where users can create compartments, interacting components of the descriptive biological system which automatically gets converted into corresponding mathematical system of ordinary or partial differential equations. Alternatively, the biological system of interest can also be specified directly in terms of mathematical equations/ expressions.

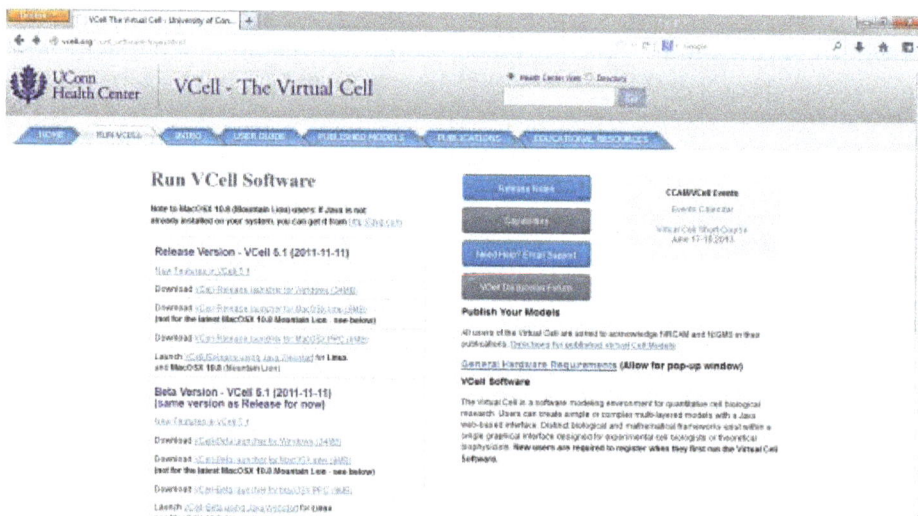

Snapshot of the webpage where the software can be downloaded

The numerical solvers (Ordinary differential Equation/ ODE solver, Partial Differential Equation/PDE solver, Parameter sensitivity analysis tool, Stochastic compartmental solver, Deterministic spatial solvers, Stochastic spatial /Smoldyn solver) of this tool then solves the equations. Direct inference can be made from the results obtained or downloaded in diverse formats. It can be downloaded from the developers website for which the link is specified below, and the snapshot is shown in Figure.

Model Development and Simulation

With the software installed, registered and connected to the virtual cell server using the registered userid and password the user can build model, perform simulations to view the output/ emerging property of cellular interaction system of their interest. Snapshot of the registered software interface is shown in Figure. This is the default view/ model and it is automatically mapped as a single compartment. The software layout is such that on the top left interface, the Bio Models consists of (i) Physiology wherein the structure, reaction diagram, reaction species are fed as inputs and the software automatically generates the mathematical equations based on the inputs provided; (ii) Applications where the numerical values of the parameters, functions and specifications are fed to set up Simulations; (iii) Pathway where in species interaction map and the network summary could be found.

All these information can seen, added, modified by clicking them. The top right interface a select ⬚, compartmental tool to add 'compartments' and species tool to add or remove 'species' directly for graphical diagram based modelling (shown in red coloured rectangle in Figure).

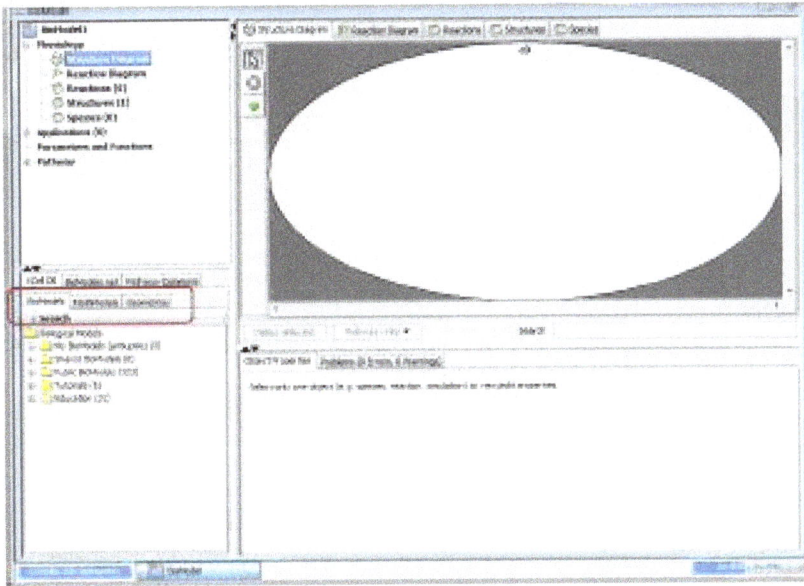

Snapshot of the registered software interface, note the bottom left corner
of the snapshot which indicates "CONNECTED (userid)".

This top right interface displays more detailed options related to Physiology, Application and Pathway. For instance, under Reaction Diagram under Physiology one can specify the relationship between the species through tools, zoom in and out of network, visualize. Similarly Reactions, Structures and Species has more options for precise representation of the interacting cellular system of interest. Once after defining the Physiology, Application can be set for a Bio Model to simulated and observe the emerging properties of the complex cellular system of interest. Application is where the systems Geometry such as size, Specifications such as initial conditions of different species of the system, Protocols, Parameter Estimation data are fed to generate mathematical model and Simulate. The three main components/layers of the software Bio Model, Geometry and Math Model are seen at the bottom left corner of the interface under VCELL DB. These three components refer to each other, however they are saved separately independent of each other. Other than the VCELL DB this software has access to the BioModels.net and Pathway Commons from where publicly available models can be accessed and/ or imported to simulate in using this tool. The lower right interface displays the object properties, problems, errors or inconsistencies and database file info.

(a)

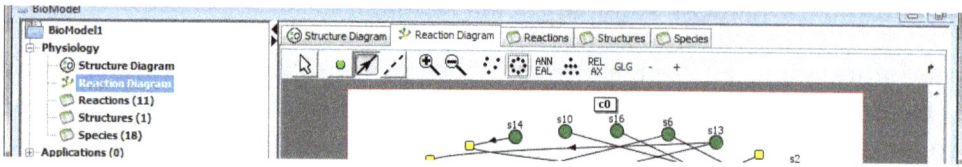

(b)

(c)

(d)

Series of snapshots showing diverse physiological representation that the Bio Model interface provides, (a) different species that the interactions among the species can be specified, this also had different visualization options to view large scale complex networks; (b) reactions between the species automatically generated based on the reaction diagram information; (c) structure indicates single compartment 'co', (d) Individual species information with the compartment in which they are present (in this case single compartment 'co').

(a)

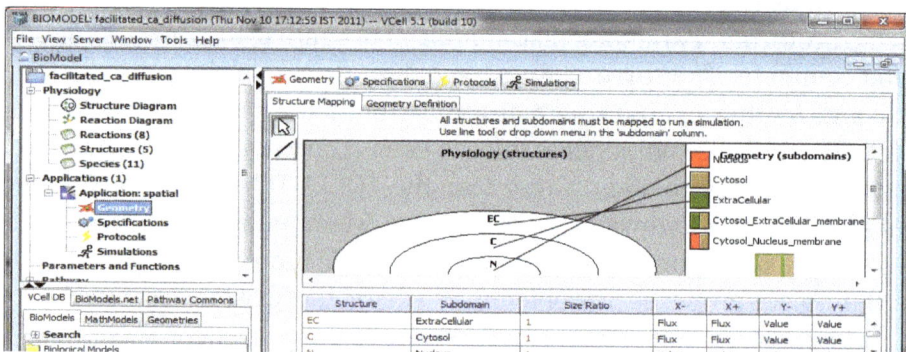

(b)

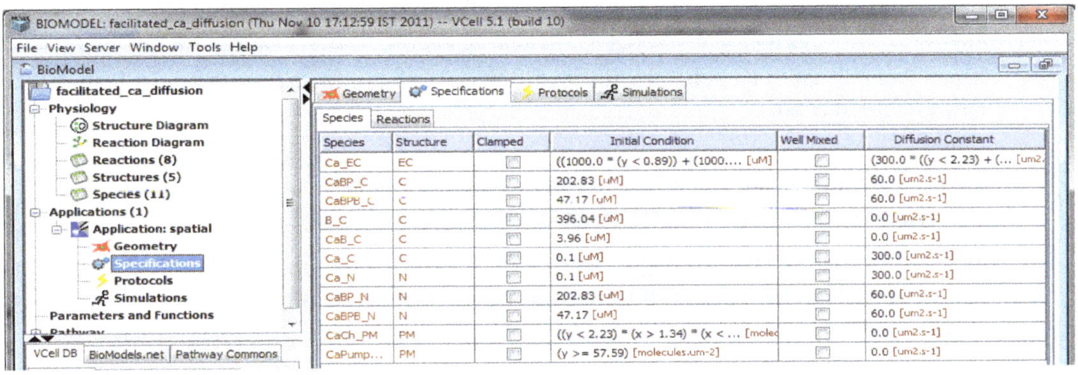

(c)

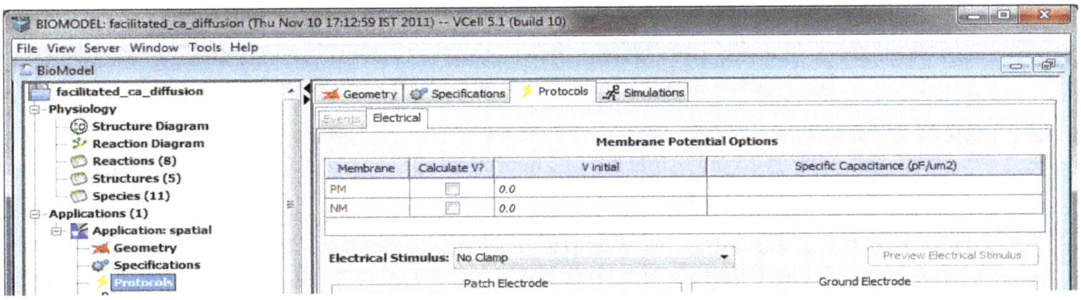

(d)

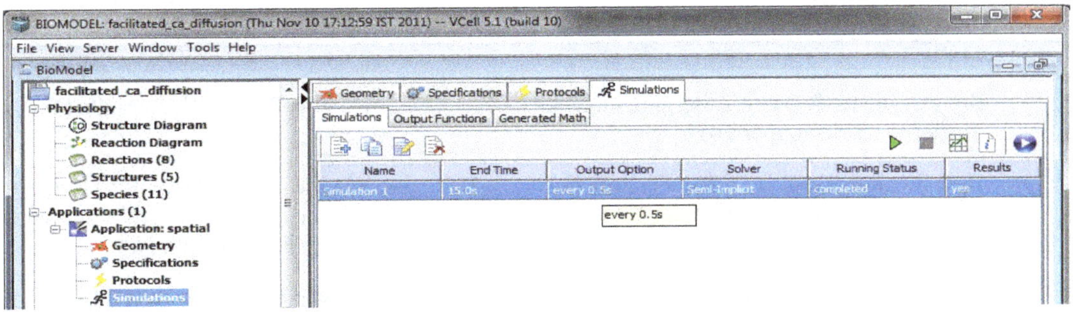

Series of snapshots showing diverse Applications (a) geometry, (b) species parameter specifications such as initial conditions and rate constants, (c) protocols, (d) Simulations details that the Bio Model interface needs for simulate the user defined system.

References

- Noble, Denis (2006). The music of life: Biology beyond the genome. Oxford: Oxford University Press. p. 176. ISBN 978-0-19-929573-9

- "Open Bioinformatics Foundation: About us". Official website. Open Bioinformatics Foundation. Retrieved 10 May 2011

- Hodgkin, Alan L; Huxley, Andrew F (28 August 1952). "A quantitative description of membrane current and its application to conduction and excitation in nerve". Journal of Physiology. 117 (4): 500–544. PMC 1392413. PMID 12991237. doi:10.1113/jphysiol.1952.sp004764

- Baitaluk, M. (2009). "System Biology of Gene Regulation". Biomedical Informatics. Methods in Molecular Biology. 569. pp. 55–87. doi:10.1007/978-1-59745-524-4_4. ISBN 978-1-934115-63-3. PMID 19623486

- Moody, Glyn (2004). Digital Code of Life: How Bioinformatics is Revolutionizing Science, Medicine, and Business. ISBN 978-0-471-32788-2

- Hogeweg P (2011). Searls, David B., ed. "The Roots of Bioinformatics in Theoretical Biology". PLoS Computational Biology. 7 (3): e1002021. Bibcode:2011PLSCB...7E0020H. PMC 3068925. PMID 21483479. doi:10.1371/journal.pcbi.1002021

- Von Bertalanffy, Ludwig (28 March 1976) [1968]. General System theory: Foundations, Development, Applications. George Braziller. p. 295. ISBN 978-0-8076-0453-3

- Kell, D. B.; Mendes, P. (2008). "The markup is the model: Reasoning about systems biology models in the Semantic Web era". Journal of Theoretical Biology. 252 (3): 538–543. doi:10.1016/j.jtbi.2007.10.023. PMID 18054049

- Nisbet, Robert (14 May 2009). "BIOINFORMATICS". Handbook of Statistical Analysis and Data Mining Applications. John Elder IV, Gary Miner. Academic Press. p. 328. Retrieved 9 May 2014

- Ionescu-Tîrgovişte, Constantin; Gagniuc, Paul Aurelian; Guja, Cristian. "Structural Properties of Gene Promoters Highlight More than Two Phenotypes of Diabetes". PLOS ONE. 10 (9): e0137950. PMC 4574929. PMID 26379145. doi:10.1371/journal.pone.0137950

- Tavassoly, Iman (2015). Dynamics of Cell Fate Decision Mediated by the Interplay of Autophagy and Apoptosis in Cancer Cells. Springer International Publishing. ISBN 978-3-319-14961-5

- Brett G. Olivier & Frank T. Bergmann. "SBML Level 3 Package: Flux Balance Constraints ('fbc')". Retrieved 24 November 2015

- Hoy, JA; Robinson, H; Trent JT, 3rd; Kakar, S; Smagghe, BJ; Hargrove, MS (3 August 2007). "Plant hemoglobins: a molecular fossil record for the evolution of oxygen transport.". Journal of Molecular Biology. 371 (1): 168–79. PMID 17560601. doi:10.1016/j.jmb.2007.05.029

- Wong, KC (2016). Computational Biology and Bioinformatics: Gene Regulation. CRC Press (Taylor & Francis Group). ISBN 9781498724975

- Ralph Gauges; Sven Sahle; Katjia Wengler; Frank T. Bergmann & Sarah M. Keating. "SBML Level 3 Package: Rendering ('render')". Retrieved 24 November 2015

- Barker, D; Alderson, R.G; McDonagh, J.L; Plaisier, H; Comrie, M.M; Duncan, L; Muirhead, G.T.P; Sweeny, S.D. (2015). "University-level practical activities in bioinformatics benefit voluntary groups of pupils in the last 2 years of school". International Journal of STEM Education. 2 (17). doi:10.1186/s40594-015-0030-z

- Kovitz, Benjamin (June 2004). "MIME Media Type for the Systems Biology Markup Language (SBML).". IETF Request for Comments 3823. Retrieved 3 January 2010

- Alm, Rebekka; Waltemath, Dagmar; Wolfien, Markus; Wolkenhauer, Olaf; Henkel, Ron (2015). "Annotation-based feature extraction from sets of SBML models". Journal of Biomedical Semantics. 6. doi:10.1186/s13326-015-0014-4

- Encyclopedia of Systems Biology Dubitzky, W., Wolkenhauer, O., Yokota, H., Cho, K.-H. (Eds.) SBML, pp2057-2062 Springer 2013 ISBN 978-1-4419-9863-7

Gene Expression and its Networks

The information that is stored in our genes can be transferred into a functional product. One of the examples of these functional products is protein. Some of the techniques of measuring gene expression are reporter gene, western blot, SAGE, DNA microarray and RNA-Seq. The section serves as a source to understand the major categories related to gene expression.

Gene Expression

Gene expression is the process by which information from a gene is used in the synthesis of a functional gene product. These products are often proteins, but in non-protein coding genes such as transfer RNA (tRNA) or small nuclear RNA (snRNA) genes, the product is a functional RNA. The process of gene expression is used by all known life—eukaryotes (including multicellular organisms), prokaryotes (bacteria and archaea), and utilized by viruses—to generate the macromolecular machinery for life.

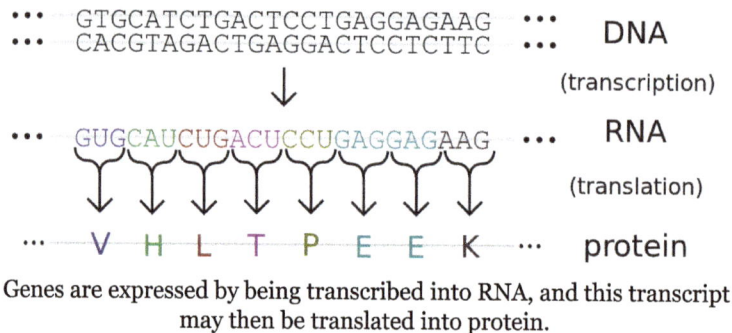

```
···   GTGCATCTGACTCCTGAGGAGAAG  ···   DNA
···   CACGTAGACTGAGGACTCCTCTTC  ···
                                      (transcription)
                  ↓
···   GUGCAUCUGACUCCUGAGGAGAAG  ···   RNA
      ⌣ ⌣ ⌣ ⌣ ⌣ ⌣ ⌣ ⌣
      ↓ ↓ ↓ ↓ ↓ ↓ ↓ ↓              (translation)
···   V  H  L  T  P  E  E  K  ···   protein
```

Genes are expressed by being transcribed into RNA, and this transcript may then be translated into protein.

Several steps in the gene expression process may be modulated, including the transcription, RNA splicing, translation, and post-translational modification of a protein. Gene regulation gives the cell control over structure and function, and is the basis for cellular differentiation, morphogenesis and the versatility and adaptability of any organism. Gene regulation may also serve as a substrate for evolutionary change, since control of the timing, location, and amount of gene expression can have a profound effect on the functions (actions) of the gene in a cell or in a multicellular organism.

In genetics, gene expression is the most fundamental level at which the genotype gives rise to the phenotype, i.e. observable trait. The genetic code stored in DNA is "interpreted" by gene expression, and the properties of the expression give rise to the organism's phenotype. Such phenotypes are often expressed by the synthesis of proteins that control the organism's shape, or that act as enzymes catalysing specific metabolic pathways characterising the organism. Regulation of gene expression is thus critical to an organism's development.

Mechanism

Transcription

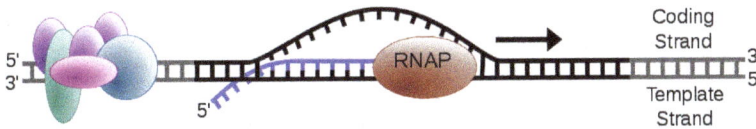

The process of transcription is carried out by RNA polymerase (RNAP), which uses DNA (black) as a template and produces RNA (blue).

A gene is a stretch of DNA that encodes information. Genomic DNA consists of two antiparallel and reverse complementary strands, each having 5' and 3' ends. With respect to a gene, the two strands may be labeled the "template strand," which serves as a blueprint for the production of an RNA transcript, and the "coding strand," which includes the DNA version of the transcript sequence. (Perhaps surprisingly, the "coding strand" is not physically involved in the coding process because it is the "template strand" that is read during transcription.)

The production of the RNA copy of the DNA is called transcription, and is performed in the nucleus by RNA polymerase, which adds one RNA nucleotide at a time to a growing RNA strand as per the complementarity law of the bases. This RNA is complementary to the template 3' → 5' DNA strand, which is itself complementary to the coding 5' → 3' DNA strand. Therefore, the resulting 5' → 3' RNA strand is identical to the coding DNA strand with the exception that thymines (T) are replaced with uracils (U) in the RNA. A coding DNA strand reading "ATG" is indirectly transcribed through the non-coding strand as "AUG" in RNA.

In prokaryotes, transcription is carried out by a single type of RNA polymerase, which needs a DNA sequence called a Pribnow box as well as a sigma factor (σ factor) to start transcription. In eukaryotes, transcription is performed by three types of RNA polymerases, each of which needs a special DNA sequence called the promoter and a set of DNA-binding proteins—transcription factors—to initiate the process. RNA polymerase I is responsible for transcription of ribosomal RNA (rRNA) genes. RNA polymerase II (Pol II) transcribes all protein-coding genes but also some non-coding RNAs (e.g., snRNAs, snoRNAs or long non-coding RNAs). Pol II includes a C-terminal domain (CTD) that is rich in serine residues. When these residues are phosphorylated, the CTD binds to various protein factors that promote transcript maturation and modification. RNA polymerase III transcribes 5S rRNA, transfer RNA (tRNA) genes, and some small non-coding RNAs (e.g., 7SK). Transcription ends when the polymerase encounters a sequence called the terminator.

RNA Processing

While transcription of prokaryotic protein-coding genes creates messenger RNA (mRNA) that is ready for translation into protein, transcription of eukaryotic genes leaves a primary transcript of RNA (pre-mRNA), which first has to undergo a series of modifications to become a mature mRNA.

These include 5' *capping*, which is set of enzymatic reactions that add 7-methylguanosine (m^7G) to the 5' end of pre-mRNA and thus protect the RNA from degradation by exonucleases. The m^7G cap is then bound by cap binding complex heterodimer (CBC20/CBC80), which aids in mRNA export to cytoplasm and also protect the RNA from decapping.

Another modification is 3' *cleavage and polyadenylation*. They occur if polyadenylation signal sequence (5'- AAUAAA-3') is present in pre-mRNA, which is usually between protein-coding sequence and terminator. The pre-mRNA is first cleaved and then a series of ~200 adenines (A) are added to form poly(A) tail, which protects the RNA from degradation. Poly(A) tail is bound by multiple poly(A)-binding proteins (PABP) necessary for mRNA export and translation re-initiation.

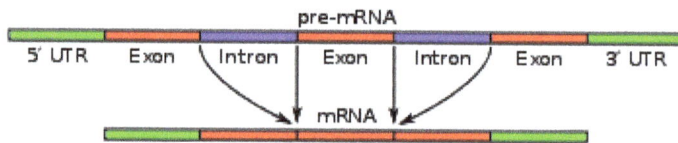

Simple illustration of exons and introns in pre-mRNA and the formation of mature mRNA by splicing. The UTRs are non-coding parts of exons at the ends of the mRNA.

A very important modification of eukaryotic pre-mRNA is *RNA splicing*. The majority of eukaryotic pre-mRNAs consist of alternating segments called exons and introns. During the process of splicing, an RNA-protein catalytical complex known as spliceosome catalyzes two transesterification reactions, which remove an intron and release it in form of lariat structure, and then splice neighbouring exons together. In certain cases, some introns or exons can be either removed or retained in mature mRNA. This so-called alternative splicing creates series of different transcripts originating from a single gene. Because these transcripts can be potentially translated into different proteins, splicing extends the complexity of eukaryotic gene expression.

Extensive RNA processing may be an evolutionary advantage made possible by the nucleus of eukaryotes. In prokaryotes, transcription and translation happen together, whilst in eukaryotes, the nuclear membrane separates the two processes, giving time for RNA processing to occur.

Non-coding RNA Maturation

In most organisms non-coding genes (ncRNA) are transcribed as precursors that undergo further processing. In the case of ribosomal RNAs (rRNA), they are often transcribed as a pre-rRNA that contains one or more rRNAs. The pre-rRNA is cleaved and modified (2'-O-methylation and pseudouridine formation) at specific sites by approximately 150 different small nucleolus-restricted RNA species, called snoRNAs. SnoRNAs associate with proteins, forming snoRNPs. While snoRNA part basepair with the target RNA and thus position the modification at a precise site, the protein part performs the catalytical reaction. In eukaryotes, in particular a snoRNP called RNase, MRP cleaves the 45S pre-rRNA into the 28S, 5.8S, and 18S rRNAs. The rRNA and RNA processing factors form large aggregates called the nucleolus.

In the case of transfer RNA (tRNA), for example, the 5' sequence is removed by RNase P, whereas the 3' end is removed by the tRNase Z enzyme and the non-templated 3' CCA tail is added by a nucleotidyl transferase. In the case of micro RNA (miRNA), miRNAs are first transcribed as primary transcripts or pri-miRNA with a cap and poly-A tail and processed to short, 70-nucleotide stem-loop structures known as pre-miRNA in the cell nucleus by the enzymes Drosha and Pasha. After being exported, it is then processed to mature miRNAs in the cytoplasm by interaction with the endonuclease Dicer, which also initiates the formation of the RNA-induced silencing complex (RISC), composed of the Argonaute protein.

Even snRNAs and snoRNAs themselves undergo series of modification before they become part of functional RNP complex. This is done either in the nucleoplasm or in the specialized compartments called Cajal bodies. Their bases are methylated or pseudouridinilated by a group of small Cajal body-specific RNAs (scaRNAs), which are structurally similar to snoRNAs.

RNA Export

In eukaryotes most mature RNA must be exported to the cytoplasm from the nucleus. While some RNAs function in the nucleus, many RNAs are transported through the nuclear pores and into the cytosol. Notably this includes all RNA types involved in protein synthesis. In some cases RNAs are additionally transported to a specific part of the cytoplasm, such as a synapse; they are then towed by motor proteins that bind through linker proteins to specific sequences (called "zipcodes") on the RNA.

Translation

For some RNA (non-coding RNA) the mature RNA is the final gene product. In the case of messenger RNA (mRNA) the RNA is an information carrier coding for the synthesis of one or more proteins. mRNA carrying a single protein sequence (common in eukaryotes) is monocistronic whilst mRNA carrying multiple protein sequences (common in prokaryotes) is known as polycistronic.

Every mRNA consists of three parts: a 5' untranslated region (5'UTR), a protein-coding region or open reading frame (ORF), and a 3' untranslated region (3'UTR). The coding region carries information for protein synthesis encoded by the genetic code to form triplets. Each triplet of nucleotides of the coding region is called a codon and corresponds to a binding site complementary to an anticodon triplet in transfer RNA. Transfer RNAs with the same anticodon sequence always carry an identical type of amino acid. Amino acids are then chained together by the ribosome according to the order of triplets in the coding region. The ribosome helps transfer RNA to bind to messenger RNA and takes the amino acid from each transfer RNA and makes a structure-less protein out of it. Each mRNA molecule is translated into many protein molecules, on average ~2800 in mammals.

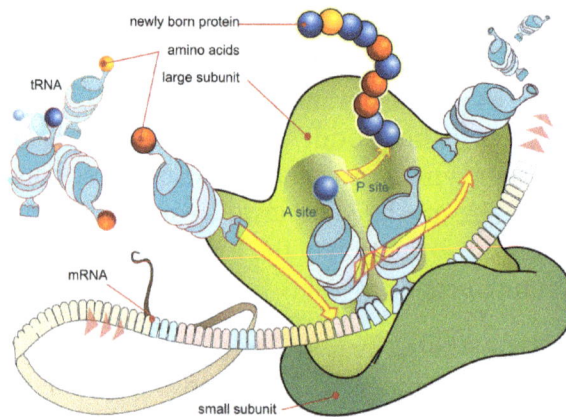

During the translation, tRNA charged with amino acid enters the ribosome and aligns with the correct mRNA triplet. Ribosome then adds amino acid to growing protein chain.

In prokaryotes translation generally occurs at the point of transcription (co-transcriptionally), often using a messenger RNA that is still in the process of being created. In eukaryotes translation can occur in a variety of regions of the cell depending on where the protein being written is supposed to be. Major locations are the cytoplasm for soluble cytoplasmic proteins and the membrane of the endoplasmic reticulum for proteins that are for export from the cell or insertion into a cell membrane. Proteins that are supposed to be expressed at the endoplasmic reticulum are recognised part-way through the translation process. This is governed by the signal recognition particle—a protein that binds to the ribosome and directs it to the endoplasmic reticulum when it finds a signal peptide on the growing (nascent) amino acid chain. Translation is the communication of the meaning of a source-language text by means of an equivalent target-language text

Folding

The polypeptide folds into its characteristic and functional three-dimensional structure from a random coil. Each protein exists as an unfolded polypeptide or random coil when translated from a sequence of mRNA into a linear chain of amino acids. This polypeptide lacks any developed three-dimensional structure (the left hand side of the neighboring figure). Amino acids interact with each other to produce a well-defined three-dimensional structure, the folded protein (the right hand side of the figure) known as the native state. The resulting three-dimensional structure is determined by the amino acid sequence (Anfinsen's dogma).

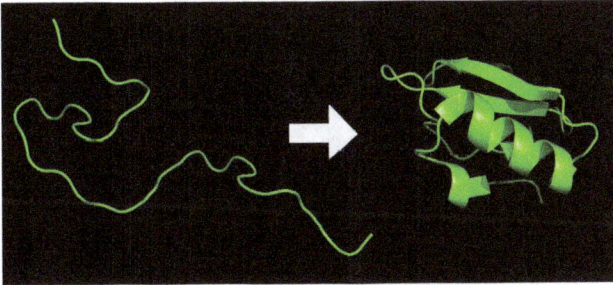

Protein before (left) and after (right) folding.

The correct three-dimensional structure is essential to function, although some parts of functional proteins may remain unfolded Failure to fold into the intended shape usually produces inactive proteins with different properties including toxic prions. Several neurodegenerative and other diseases are believed to result from the accumulation of *misfolded* proteins. Many allergies are caused by the folding of the proteins, for the immune system does not produce antibodies for certain protein structures.

Enzymes called chaperones assist the newly formed protein to attain (fold into) the 3-dimensional structure it needs to function. Similarly, RNA chaperones help RNAs attain their functional shapes. Assisting protein folding is one of the main roles of the endoplasmic reticulum in eukaryotes.

Translocation

Secretory proteins of eukaryotes or prokaryotes must be translocated to enter the secretory pathway. Newly synthesized proteins are directed to the eukaryotic Sec61 or prokaryotic SecYEG translocation channel by signal peptides. The efficiency of protein secretion in eukaryotes is very dependent on the signal peptide which has been used.

Protein transport

Many proteins are destined for other parts of the cell than the cytosol and a wide range of signalling sequences or (signal peptides) are used to direct proteins to where they are supposed to be. In prokaryotes this is normally a simple process due to limited compartmentalisation of the cell. However, in eukaryotes there is a great variety of different targeting processes to ensure the protein arrives at the correct organelle.

Not all proteins remain within the cell and many are exported, for example, digestive enzymes, hormones and extracellular matrix proteins. In eukaryotes the export pathway is well developed and the main mechanism for the export of these proteins is translocation to the endoplasmic reticulum, followed by transport via the Golgi apparatus.

Regulation of Gene Expression

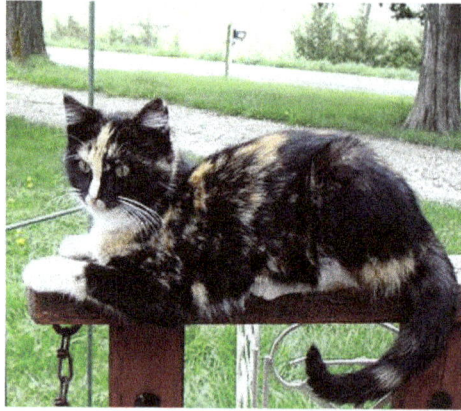

The patchy colours of a tortoiseshell cat are the result of different levels of expression
of pigmentation genes in different areas of the skin.

Regulation of gene expression refers to the control of the amount and timing of appearance of the functional product of a gene. Control of expression is vital to allow a cell to produce the gene products it needs when it needs them; in turn, this gives cells the flexibility to adapt to a variable environment, external signals, damage to the cell, and other stimuli. More generally, gene regulation gives the cell control over all structure and function, and is the basis for cellular differentiation, morphogenesis and the versatility and adaptability of any organism.

Numerous terms are used to describe types of genes depending on how they are regulated; these include:

- A constitutive gene is a gene that is transcribed continually as opposed to a facultative gene, which is only transcribed when needed.

- A *housekeeping gene* is a gene that is required to maintain basic cellular function and so is typically expressed in all cell types of an organism. Examples include actin, GAPDH and ubiquitin. Some housekeeping genes are transcribed at a relatively constant rate and these genes can be used as a reference point in experiments to measure the expression rates of other genes.

- A facultative gene is a gene only transcribed when needed as opposed to a constitutive gene.

- An inducible gene is a gene whose expression is either responsive to environmental change or dependent on the position in the cell cycle.

Any step of gene expression may be modulated, from the DNA-RNA transcription step to post-translational modification of a protein. The stability of the final gene product, whether it is RNA or protein, also contributes to the expression level of the gene—an unstable product results in a low expression level. In general gene expression is regulated through changes in the number and type of interactions between molecules that collectively influence transcription of DNA and translation of RNA.

Some simple examples of where gene expression is important are:

- Control of insulin expression so it gives a signal for blood glucose regulation.

- X chromosome inactivation in female mammals to prevent an "overdose" of the genes it contains.

- Cyclin expression levels control progression through the eukaryotic cell cycle.

Transcriptional Regulation

Regulation of transcription can be broken down into three main routes of influence; genetic (direct interaction of a control factor with the gene), modulation interaction of a control factor with the transcription machinery and epigenetic (non-sequence changes in DNA structure that influence transcription).

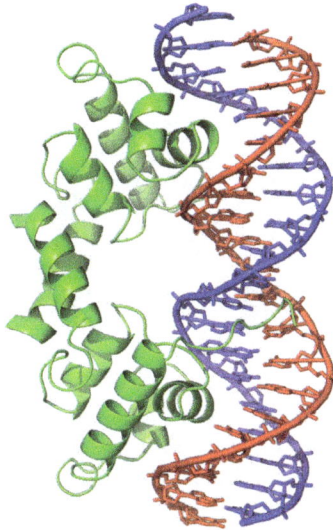

The lambda repressor transcription factor (green) binds as a dimer to major groove of DNA target (red and blue) and disables initiation of transcription. From PDB: 1LMB.

Direct interaction with DNA is the simplest and the most direct method by which a protein changes transcription levels. Genes often have several protein binding sites around the coding region with the specific function of regulating transcription. There are many classes of regulatory DNA binding sites known as enhancers, insulators and silencers. The mechanisms for regulating transcription are very varied, from blocking key binding sites on the DNA for RNA polymerase to acting as an activator and promoting transcription by assisting RNA polymerase binding.

The activity of transcription factors is further modulated by intracellular signals causing protein post-translational modification including phosphorylated, acetylated, or glycosylated. These changes influence a transcription factor's ability to bind, directly or indirectly, to promoter DNA, to recruit RNA polymerase, or to favor elongation of a newly synthesized RNA molecule.

The nuclear membrane in eukaryotes allows further regulation of transcription factors by the duration of their presence in the nucleus, which is regulated by reversible changes in their structure and by binding of other proteins. Environmental stimuli or endocrine signals may cause modification of regulatory proteins eliciting cascades of intracellular signals, which result in regulation of gene expression.

More recently it has become apparent that there is a significant influence of non-DNA-sequence specific effects on transcription. These effects are referred to as epigenetic and involve the higher order structure of DNA, non-sequence specific DNA binding proteins and chemical modification of DNA. In general epigenetic effects alter the accessibility of DNA to proteins and so modulate transcription.

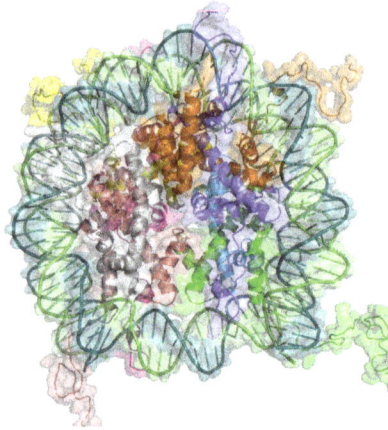

In eukaryotes, DNA is organized in form of nucleosomes. Note how the DNA (blue and green) is tightly wrapped around the protein core made of histone octamer (ribbon coils), restricting access to the DNA. From PDB: 1KX5.

DNA methylation is a widespread mechanism for epigenetic influence on gene expression and is seen in bacteria and eukaryotes and has roles in heritable transcription silencing and transcription regulation. In eukaryotes the structure of chromatin, controlled by the histone code, regulates access to DNA with significant impacts on the expression of genes in euchromatin and heterochromatin areas.

Transcriptional Regulation in Cancer

The majority of gene promoters contain a CpG island with numerous CpG sites. When many of a gene's promoter CpG sites are methylated the gene becomes silenced. Colorectal cancers typically have 3 to 6 driver mutations and 33 to 66 hitchhiker or passenger mutations. However, transcriptional silencing may be of more importance than mutation in causing progression to cancer. For example, in colorectal cancers about 600 to 800 genes are transcriptionally silenced by CpG island methylation. Transcriptional repression in cancer can also occur by other epigenetic mechanisms, such as altered expression of microRNAs. In breast cancer, transcriptional repression of BRCA1 may occur more frequently by over-expressed microRNA-182 than by hypermethylation of the BRCA1 promoter.

Post-transcriptional Regulation

In eukaryotes, where export of RNA is required before translation is possible, nuclear export is thought to provide additional control over gene expression. All transport in and out of the nucleus is via the nuclear pore and transport is controlled by a wide range of importin and exportin proteins.

Expression of a gene coding for a protein is only possible if the messenger RNA carrying the code survives long enough to be translated. In a typical cell, an RNA molecule is only stable if specifically protected from degradation. RNA degradation has particular importance in regulation of expression in eukaryotic cells where mRNA has to travel significant distances before being translated. In eukaryotes, RNA is stabilised by certain post-transcriptional modifications, particularly the 5' cap and poly-adenylated tail.

Intentional degradation of mRNA is used not just as a defence mechanism from foreign RNA (normally from viruses) but also as a route of mRNA *destabilisation*. If an mRNA molecule has a complementary sequence to a small interfering RNA then it is targeted for destruction via the RNA interference pathway.

Three Prime Untranslated Regions and MicroRNAs

Three prime untranslated regions (3'UTRs) of messenger RNAs (mRNAs) often contain regulatory sequences that post-transcriptionally influence gene expression. Such 3'-UTRs often contain both binding sites for microRNAs (miRNAs) as well as for regulatory proteins. By binding to specific sites within the 3'-UTR, miRNAs can decrease gene expression of various mRNAs by either inhibiting translation or directly causing degradation of the transcript. The 3'-UTR also may have silencer regions that bind repressor proteins that inhibit the expression of a mRNA.

The 3'-UTR often contains microRNA response elements (MREs). MREs are sequences to which miRNAs bind. These are prevalent motifs within 3'-UTRs. Among all regulatory motifs within the 3'-UTRs (e.g. including silencer regions), MREs make up about half of the motifs.

As of 2014, the miRBase web site, an archive of miRNA sequences and annotations, listed 28,645 entries in 233 biologic species. Of these, 1,881 miRNAs were in annotated human miRNA loci. miRNAs were predicted to have an average of about four hundred target mRNAs (affecting expression of several hundred genes). Freidman et al.

estimate that >45,000 miRNA target sites within human mRNA 3'UTRs are conserved above background levels, and >60% of human protein-coding genes have been under selective pressure to maintain pairing to miRNAs.

Direct experiments show that a single miRNA can reduce the stability of hundreds of unique mRNAs. Other experiments show that a single miRNA may repress the production of hundreds of proteins, but that this repression often is relatively mild (less than 2-fold).

The effects of miRNA dysregulation of gene expression seem to be important in cancer. For instance, in gastrointestinal cancers, nine miRNAs have been identified as epigenetically altered and effective in down regulating DNA repair enzymes.

The effects of miRNA dysregulation of gene expression also seem to be important in neuropsychiatric disorders, such as schizophrenia, bipolar disorder, major depression, Parkinson's disease, Alzheimer's disease and autism spectrum disorders.

Translational Regulation

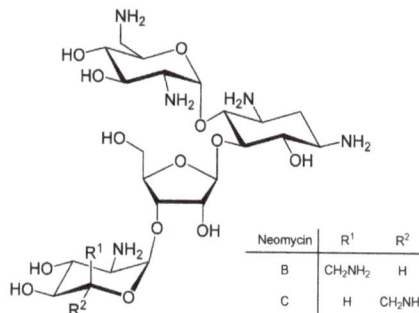

Neomycin	R^1	R^2
B	CH_2NH_2	H
C	H	CH_2NH_2

Neomycin is an example of a small molecule that reduces expression of all protein genes inevitably leading to cell death; it thus acts as an antibiotic.

Direct regulation of translation is less prevalent than control of transcription or mRNA stability but is occasionally used. Inhibition of protein translation is a major target for toxins and antibiotics, so they can kill a cell by overriding its normal gene expression control. Protein synthesis inhibitors include the antibiotic neomycin and the toxin ricin.

Protein Degradation

Once protein synthesis is complete, the level of expression of that protein can be reduced by protein degradation. There are major protein degradation pathways in all prokaryotes and eukaryotes, of which the proteasome is a common component. An unneeded or damaged protein is often labeled for degradation by addition of ubiquitin.

Measurement

Measuring gene expression is an important part of many life sciences, as the ability to quantify the level at which a particular gene is expressed within a cell, tissue or organism can provide a lot of valuable information. For example, measuring gene expression can:

- Identify viral infection of a cell (viral protein expression).

- Determine an individual's susceptibility to cancer (oncogene expression).

- Find if a bacterium is resistant to penicillin (beta-lactamase expression).

Similarly, the analysis of the location of protein expression is a powerful tool, and this can be done on an organismal or cellular scale. Investigation of localization is particularly important for the study of development in multicellular organisms and as an indicator of protein function in single cells. Ideally, measurement of expression is done by detecting the final gene product (for many genes, this is the protein); however, it is often easier to detect one of the precursors, typically mRNA and to infer gene-expression levels from these measurements.

mRNA quantification

Levels of mRNA can be quantitatively measured by northern blotting, which provides size and sequence information about the mRNA molecules. A sample of RNA is separated on an agarose gel and hybridized to a radioactively labeled RNA probe that is complementary to the target sequence. The radiolabeled RNA is then detected by an autoradiograph. Because the use of radioactive reagents makes the procedure time consuming and potentially dangerous, alternative labeling and detection methods, such as digoxigenin and biotin chemistries, have been developed. Perceived disadvantages of Northern blotting are that large quantities of RNA are required and that quantification may not be completely accurate, as it involves measuring band strength in an image of a gel. On the other hand, the additional mRNA size information from the Northern blot allows the discrimination of alternately spliced transcripts.

Another approach for measuring mRNA abundance is RT-qPCR. In this technique, reverse transcription is followed by quantitative PCR. Reverse transcription first generates a DNA template from the mRNA; this single-stranded template is called cDNA. The cDNA template is then amplified in the quantitative step, during which the fluorescence emitted by labeled hybridization probes or intercalating dyes changes as the DNA amplification process progresses. With a carefully constructed standard curve, qPCR can produce an absolute measurement of the number of copies of original mRNA, typically in units of copies per nanolitre of homogenized tissue or copies per cell. qPCR is very sensitive (detection of a single mRNA molecule is theoretically possible), but can be expensive depending on the type of reporter used; fluorescently labeled oligonucleotide probes are more expensive than non-specific intercalating fluorescent dyes.

For expression profiling, or high-throughput analysis of many genes within a sample, quantitative PCR may be performed for hundreds of genes simultaneously in the case of low-density arrays. A second approach is the hybridization microarray. A single array or "chip" may contain probes to determine transcript levels for every known gene in the genome of one or more organisms. Alternatively, "tag based" technologies like Serial analysis of gene expression (SAGE) and RNA-Seq, which can provide a relative measure of the cellular concentration of different mRNAs, can be used. An advantage of tag-based methods is the "open architecture", allowing for the exact measurement of any transcript, with a known or unknown sequence. Next-generation sequencing (NGS) such as RNA-Seq is another approach, producing vast quantities of sequence data that can be matched to a reference genome. Although NGS is comparatively time-consuming, expensive, and resource-intensive, it can identify single-nucleotide polymorphisms, splice-variants, and novel genes, and can also be used to profile expression in organisms for which little or no sequence information is available.

Protein Quantification

For genes encoding proteins, the expression level can be directly assessed by a number of methods with some clear analogies to the techniques for mRNA quantification.

The most commonly used method is to perform a Western blot against the protein of interest—this gives information on the size of the protein in addition to its identity. A sample (often cellular lysate) is separated on a polyacrylamide gel, transferred to a membrane and then probed with an antibody to the protein of interest. The antibody can either be conjugated to a fluorophore or to horseradish peroxidase for imaging and/or quantification. The gel-based nature of this assay makes quantification less accurate, but it has the advantage of being able to identify later modifications to the protein, for example proteolysis or ubiquitination, from changes in size.

Localisation

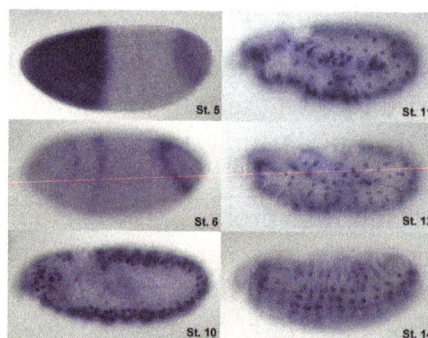

In situ-hybridization of Drosophila embryos at different developmental stages for the mRNA responsible for the expression of hunchback. High intensity of blue color marks places with high hunchback mRNA quantity.

Analysis of expression is not limited to quantification; localisation can also be determined. mRNA can be detected with a suitably labelled complementary mRNA strand

and protein can be detected via labelled antibodies. The probed sample is then observed by microscopy to identify where the mRNA or protein is.

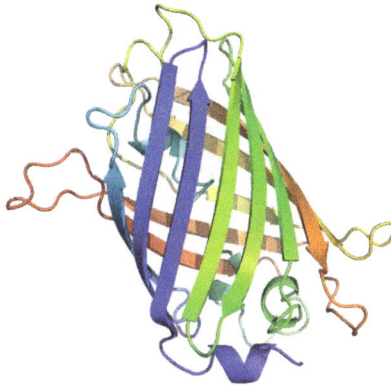

The three-dimensional structure of green fluorescent protein. The residues in the centre of the "barrel" are responsible for production of green light after exposing to higher energetic blue light. From PDB: 1EMA.

By replacing the gene with a new version fused to a green fluorescent protein (or similar) marker, expression may be directly quantified in live cells. This is done by imaging using a fluorescence microscope. It is very difficult to clone a GFP-fused protein into its native location in the genome without affecting expression levels so this method often cannot be used to measure endogenous gene expression. It is, however, widely used to measure the expression of a gene artificially introduced into the cell, for example via an expression vector. It is important to note that by fusing a target protein to a fluorescent reporter the protein's behavior, including its cellular localization and expression level, can be significantly changed.

The enzyme-linked immunosorbent assay works by using antibodies immobilised on a microtiter plate to capture proteins of interest from samples added to the well. Using a detection antibody conjugated to an enzyme or fluorophore the quantity of bound protein can be accurately measured by fluorometric or colourimetric detection. The detection process is very similar to that of a Western blot, but by avoiding the gel steps more accurate quantification can be achieved.

Expression System

An expression system is a system specifically designed for the production of a gene product of choice. This is normally a protein although may also be RNA, such as tRNA or a ribozyme. An expression system consists of a gene, normally encoded by DNA, and the molecular machinery required to transcribe the DNA into mRNA and translate the mRNA into protein using the reagents provided. In the broadest sense this includes every living cell but the term is more normally used to refer to expression as a laboratory tool. An expression system is therefore often artificial in some manner. Expression systems are, however, a fundamentally natural process. Viruses are an excellent example where they replicate by using the host cell as an expression system for the viral proteins and genome.

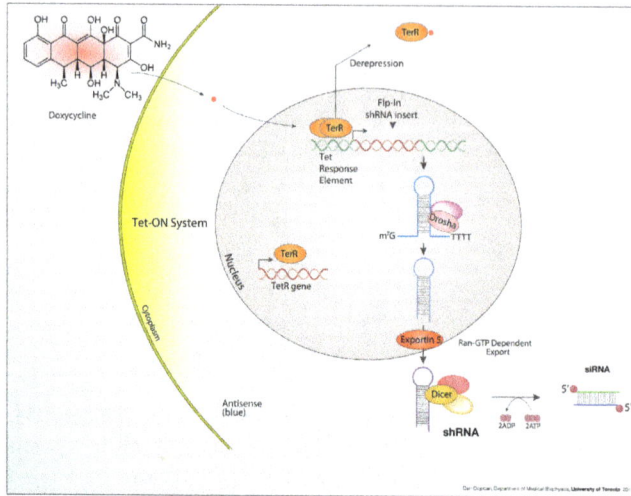

Tet-ON inducible shRNA system

Inducible Expression

Doxycycline is also used in "Tet-on" and "Tet-off" tetracycline controlled transcriptional activation to regulate transgene expression in organisms and cell cultures.

In Nature

In addition to these biological tools, certain naturally observed configurations of DNA (genes, promoters, enhancers, repressors) and the associated machinery itself are referred to as an expression system. This term is normally used in the case where a gene or set of genes is switched on under well defined conditions, for example, the simple repressor switch expression system in Lambda phage and the lac operator system in bacteria. Several natural expression systems are directly used or modified and used for artificial expression systems such as the Tet-on and Tet-off expression system.

Gene Networks

Genes have sometimes been regarded as nodes in a network, with inputs being proteins such as transcription factors, and outputs being the level of gene expression. The node itself performs a function, and the operation of these functions have been interpreted as performing a kind of information processing within cells and determines cellular behavior.

Gene networks can also be constructed without formulating an explicit causal model. This is often the case when assembling networks from large expression data sets. Co-variation and correlation of expression is computed across a large sample of cases and measurements (often transcriptome or proteome data). The source of variation can be either experimental or natural (observational). There are several ways to construct gene expression networks, but one common approach is to compute a matrix of all pair-wise correlations of expression across conditions, time points, or individuals and convert the matrix (after thresholding at some cut-off value) into a graphical representation in which nodes represent genes, transcripts, or proteins and edges connecting these nodes represent the strength of association. Weighted correlation network analysis involves weighted networks defined by soft-thresholding the pairwise correlations among variables (e.g. measures of transcript abundance).

Techniques and Tools

The following experimental techniques are used to measure gene expression and are listed in roughly chronological order, starting with the older, more established technologies. They are divided into two groups based on their degree of multiplexity.

- Low-to-mid-plex techniques:
    - Reporter gene
    - Northern blot
    - Western blot
    - Fluorescent in situ hybridization
    - Reverse transcription PCR
- Higher-plex techniques:
    - SAGE
    - DNA microarray
    - Tiling array
    - RNA-Seq

Gene Regulatory Network

A gene (or genetic) regulatory network (GRN) is a collection of molecular regulators that interact with each other and with other substances in the cell to govern the gene expression levels of mRNA and proteins. These play a central role in morphogenesis, the creation of body structures, which in turn is central to evolutionary developmental biology (evo-devo).

The regulator can be DNA, RNA, protein and complexes of these. The interaction can be direct or indirect (through transcribed RNA or translated protein). In general, each mRNA molecule goes on to make a specific protein (or set of proteins). In some cases this protein will be structural, and will accumulate at the cell membrane or within the cell to give it particular structural properties. In other cases the protein will be an enzyme, i.e., a micro-machine that catalyses a certain reaction, such as the breakdown of a food source or toxin. Some proteins though serve only to activate other genes, and these are the transcription factors that are the main players in regulatory networks or cascades. By binding to the promoter region at the start of other genes they turn them on, initiating the production of another protein, and so on. Some transcription factors are inhibitory.

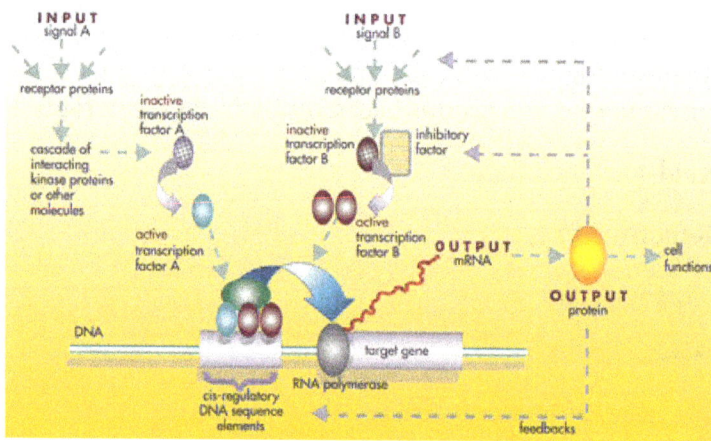

Structure of a gene regulatory network

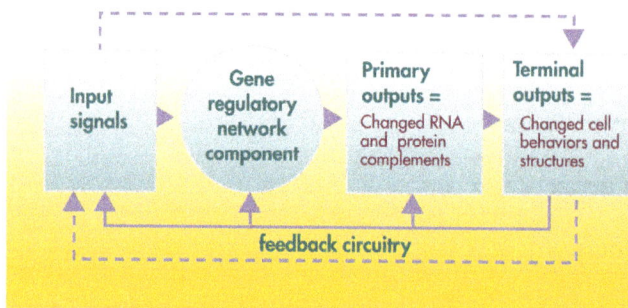

Control process of a gene regulatory network

In single-celled organisms, regulatory networks respond to the external environment, optimising the cell at a given time for survival in this environment. Thus a yeast cell, finding itself in a sugar solution, will turn on genes to make enzymes that process the sugar to alcohol. This process, which we associate with wine-making, is how the yeast cell makes its living, gaining energy to multiply, which under normal circumstances would enhance its survival prospects.

In multicellular animals the same principle has been put in the service of gene cascades that control body-shape. Each time a cell divides, two cells result which, although they contain the same genome in full, can differ in which genes are turned on and making proteins. Sometimes a 'self-sustaining feedback loop' ensures that a cell maintains its identity and passes it on. Less understood is the mechanism of epigenetics by which chromatin modification may provide cellular memory by blocking or allowing transcription. A major feature of multicellular animals is the use of morphogen gradients, which in effect provide a positioning system that tells a cell where in the body it is, and hence what sort of cell to become. A gene that is turned on in one cell may make a product that leaves the cell and diffuses through adjacent cells, entering them and turning on genes only when it is present above a certain threshold level. These cells are thus induced into a new fate, and may even generate other morphogens that signal back to the original cell. Over longer distances morphogens may use the active process of signal transduction. Such signalling controls embryogenesis, the building of a body plan from scratch through a series of sequential steps. They also control and maintain adult bodies through feedback processes, and the loss of such feedback because of a mutation can be responsible for the cell proliferation that is seen in cancer. In parallel with this process of building structure, the gene cascade turns on genes that make structural proteins that give each cell the physical properties it needs.

Overview

At one level, biological cells can be thought of as "partially mixed bags" of biological chemicals – in the discussion of gene regulatory networks, these chemicals are mostly the messenger RNAs (mRNAs) and proteins that arise from gene expression. These mRNA and proteins interact with each other with various degrees of specificity. Some diffuse around the cell. Others are bound to cell membranes, interacting with molecules in the environment. Still others pass through cell membranes and mediate long range signals to other cells in a multi-cellular organism. These molecules and their interactions comprise a *gene regulatory network*. A typical gene regulatory network looks something like this:

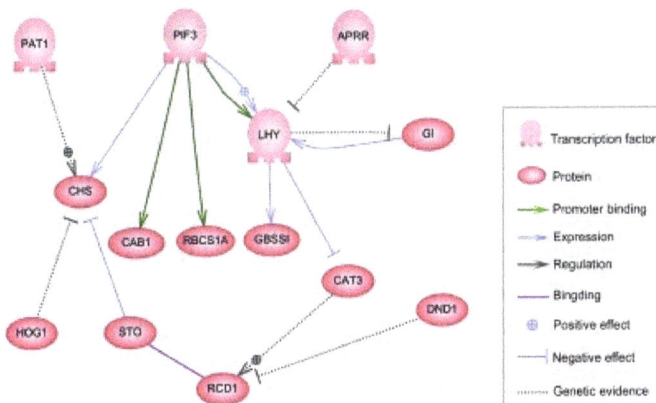

Example of a regulatory network

The nodes of this network are proteins, their corresponding mRNAs, and protein/protein complexes. Nodes that are depicted as lying along vertical lines are associated with the cell/environment interfaces, while the others are free-floating and can diffuse. Implied are genes, the DNA sequences which are transcribed into the mRNAs that translate into proteins. Edges between nodes represent individual molecular reactions, the protein/protein and protein/mRNA interactions through which the products of one gene affect those of another, though the lack of experimentally obtained information often implies that some reactions are not modeled at such a fine level of detail. These interactions can be inductive (the arrowheads), with an increase in the concentration of one leading to an increase in the other, or inhibitory (the filled circles), with an increase in one leading to a decrease in the other. A series of edges indicates a chain of such dependencies, with cycles corresponding to feedback loops. The network structure is an abstraction of the system's chemical dynamics, describing the manifold ways in which one substance affects all the others to which it is connected. In practice, such GRNs are inferred from the biological literature on a given system and represent a distillation of the collective knowledge about a set of related biochemical reactions. To speed up the manual curation of GRNs, some recent efforts try to use text mining and information extraction technologies for this purpose.

Genes can be viewed as nodes in the network, with input being proteins such as transcription factors, and outputs being the level of gene expression. The node itself can also be viewed as a function which can be obtained by combining basic functions upon the inputs (in the Boolean network described below these are Boolean functions, typically AND, OR, and NOT). These functions have been interpreted as performing a kind of information processing within the cell, which determines cellular behavior. The basic drivers within cells are concentrations of some proteins, which determine both spatial (location within the cell or tissue) and temporal (cell cycle or developmental stage) coordinates of the cell, as a kind of "cellular memory". The gene networks are only beginning to be understood, and it is a next step for biology to attempt to deduce the functions for each gene "node", to help understand the behavior of the system in increasing levels of complexity, from gene to signaling pathway, cell or tissue level.

Mathematical models of GRNs have been developed to capture the behavior of the system being modeled, and in some cases generate predictions corresponding with experimental observations. In some other cases, models have proven to make accurate novel predictions, which can be tested experimentally, thus suggesting new approaches to explore in an experiment that sometimes wouldn't be considered in the design of the protocol of an experimental laboratory. The most common modeling technique involves the use of coupled ordinary differential equations (ODEs). Several other promising modeling techniques have been used, including Boolean networks, Petri nets, Bayesian networks, graphical Gaussian models, Stochastic, and Process Calculi. Conversely, techniques have been proposed for generating models of GRNs that best explain a set of time series observations. Recently it has been shown that ChIP-seq signal of Histone modification are more correlated with transcription factor motifs at promoters in comparison to RNA level. Hence it is proposed that time-series histone modification ChIP-seq could provide more reliable inference of gene-regulatory networks in comparison to methods based on expression levels.

Structure and Evolution

Global Feature

Gene regulatory networks are generally thought to be made up of a few highly connected nodes (hubs) and many poorly connected nodes nested within a hierarchical regulatory regime. Thus gene regulatory networks approximate a hierarchical scale free network topology. This is consistent with the view that most genes have limited pleiotropy and operate within regulatory modules. This structure is thought to evolve due to the preferential attachment of duplicated genes to more highly connected genes. Recent work has also shown that natural selection tends to favor networks with sparse connectivity.

There are primarily two ways that networks can evolve, both of which can occur simultaneously. The first is that network topology can be changed by the addition or subtraction of nodes (genes) or parts of the network (modules) may be expressed in different contexts. The *Drosophila* Hippo signaling pathway provides a good example. The Hippo signaling pathway controls both mitotic growth and post-mitotic cellular differentiation. Recently it was found that the network the Hippo signaling pathway operates in differs between these two functions which in turn changes the behavior of the Hippo signaling pathway. This suggests that the Hippo signaling pathway operates as a conserved regulatory module that can be used for multiple functions depending on context. Thus, changing network topology can allow a conserved module to serve multiple functions and alter the final output of the network. The second way networks can evolve is by changing the strength of interactions between nodes, such as how strongly a transcription factor may bind to a cis-regulatory element. Such variation in strength of network edges has been shown to underlie between species variation in vulva cell fate patterning of *Caenorhabditis* worms.

Local Feature

Another widely cited characteristic of gene regulatory network is their abundance of certain repetitive sub-networks known as network motifs. Network motifs can be regarded as repetitive topological patterns when dividing a big network into small blocks. Previous analysis found several types of motifs that appeared more often in gene regulatory networks than in randomly generated networks. As an example, one such motif is called feed-forward loops, which consist three nodes. This motif is the most abundant among all possible motifs made up of three nodes, as is shown in the gene regulatory networks of fly, nematode, and human.

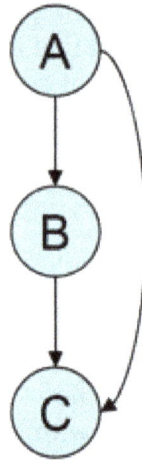

Feed-forward loop

The enriched motifs have been proposed to follow convergent evolution, suggesting they are "optimal designs" for certain regulatory purposes. For example, modeling shows that feed-forward loops are able to coordinate the change in node A (in terms of concentration and activity) and the expression dynamics of node C, creating different input-output behaviors. The galactose utilization system of *E. coli* contains a feed-forward loop which accelerates the activation of galactose utilization operon *galETK*, potentially facilitating the metabolic transition to galactose when glucose is depleted. The feed-forward loop in the arabinose utilization systems of *E.coli* delays the activation of arabinose catabolism operon and transporters, potentially avoiding unnecessary metabolic transition due to temporary fluctuations in upstream signaling pathways. Similarly in the Wnt signaling pathway of *Xenopus*, the feed-forward loop acts as a fold-change detector that responses to the fold change, rather than the absolute change, in the level of β-catenin, potentially increasing the resistance to fluctuations in β-catenin levels. Following the convergent evolution hypothesis, the enrichment of feed-forward loops would be an adaptation for fast response and noise resistance. A recent research found that yeast grown in an environment of constant glucose developed mutations in glucose signaling pathways and growth regulation pathway, suggesting regulatory components responding to environmental changes are dispensable under constant environment.

On the other hand, some researchers hypothesize that the enrichment of network motifs is non-adaptive. In other words, gene regulatory networks can evolve to a similar structure without the specific selection on the proposed input-output behavior. Support for this hypothesis often comes from computational simulations. For example, fluctuations in the abundance of feed-forward loops in a model that simulates the evolution of gene regulatory networks by randomly rewiring nodes may suggest that the enrichment of feed-forward loops is a side-effect of evolution. In another model of gene regulator networks evolution, the ratio of the frequencies of gene duplication and gene deletion show great influence on network topology: certain ratios lead to the enrichment of feed-forward loops and create networks that show features of hierarchical scale free networks.

Bacterial Regulatory Networks

Regulatory networks allow bacteria to adapt to almost every environmental niche on earth. A network of interactions among diverse types of molecules including DNA, RNA, proteins and metabolites, is utilised by the bacteria to achieve regulation of gene expression. In bacteria, the principal function of regulatory networks is to control the response to environmental changes, for example nutritional status and environmental stress. A complex organization of networks permits the microorganism to coordinate and integrate multiple environmental signals.

Modelling

Coupled Ordinary Differential Equations

It is common to model such a network with a set of coupled ordinary differential equations (ODEs) or SDEs, describing the reaction kinetics of the constituent parts. Suppose that our regulatory network has N nodes, and let $S_1(t), S_2(t), \ldots, S_N(t)$ represent the concentrations of the N corresponding substances at time t. Then the temporal evolution of the system can be described approximately by

$$\frac{dS_j}{dt} = f_j\left(S_1, S_2, \ldots, S_N\right)$$

where the functions f_j express the dependence of S_j on the concentrations of other substances present in the cell. The functions f_j are ultimately derived from basic principles of chemical kinetics or simple expressions derived from these e.g. Michaelis-Menten enzymatic kinetics. Hence, the functional forms of the f_j are usually chosen as low-order polynomials or Hill functions that serve as an ansatz for the real molecular dynamics. Such models are then studied using the mathematics of nonlinear dynamics. System-specific information, like reaction rate constants and sensitivities, are encoded as constant parameters.

By solving for the fixed point of the system:

$$\frac{dS_j}{dt} = 0$$

for all j, one obtains (possibly several) concentration profiles of proteins and mRNAs that are theoretically sustainable (though not necessarily stable). Steady states of kinetic equations thus correspond to potential cell types, and oscillatory solutions to the above equation to naturally cyclic cell types. Mathematical stability of these attractors can usually be characterized by the sign of higher derivatives at critical points, and then correspond to biochemical stability of the concentration profile. Critical points and bifurcations in the equations correspond to critical cell states in which small state or parameter perturbations could switch the system between one of several stable differentiation fates. Trajectories correspond to the unfolding of biological pathways and transients of the equations to short-term biological events.

Boolean Network

The following example illustrates how a Boolean network can model a GRN together with its gene products (the outputs) and the substances from the environment that affect it (the inputs). Stuart Kauffman was amongst the first biologists to use the metaphor of Boolean networks to model genetic regulatory networks.

1. Each gene, each input, and each output is represented by a node in a directed graph in which there is an arrow from one node to another if and only if there is a causal link between the two nodes.

2. Each node in the graph can be in one of two states: on or off.

3. For a gene, "on" corresponds to the gene being expressed; for inputs and outputs, "off" corresponds to the substance being present.

4. Time is viewed as proceeding in discrete steps. At each step, the new state of a node is a Boolean function of the prior states of the nodes with arrows pointing towards it.

The validity of the model can be tested by comparing simulation results with time series observations. A partial validation of a Boolean network model can also come from testing the predicted existence of a yet unknown regulatory connection between two particular transcription factors that each are nodes of the model.

Continuous Networks

Continuous network models of GRNs are an extension of the boolean networks described above. Nodes still represent genes and connections between them regulatory influences on gene expression. Genes in biological systems display a continuous range of activity levels and it has been argued that using a continuous representation captures several properties of gene regulatory networks not present in the Boolean model. Formally most of these approaches are similar to an artificial neural network, as inputs to a node are summed up and the result serves as input to a sigmoid function, e.g., but proteins do often control gene expression in a synergistic, i.e. non-linear, way. However, there is now a continuous network model that allows grouping of inputs to a node thus realizing another level of regulation. This model is formally closer to a higher order recurrent neural network. The same model has also been used to mimic the evolution of cellular differentiation and even multicellular morphogenesis.

Stochastic Gene Networks

Recent experimental results have demonstrated that gene expression is a stochastic process. Thus, many authors are now using the stochastic formalism, after the work by. Works on single gene expression and small synthetic genetic networks, such as the genetic toggle switch of Tim Gardner and Jim Collins, provided additional experimental data on the phenotypic variability and the stochastic nature of gene expression. The first versions of stochastic models of gene expression involved only instantaneous reactions and were driven by the Gillespie algorithm.

Since some processes, such as gene transcription, involve many reactions and could not be correctly modeled as an instantaneous reaction in a single step, it was proposed to model these reactions as single step multiple delayed reactions in order to account for the time it takes for the entire process to be complete.

From here, a set of reactions were proposed that allow generating GRNs. These are then simulated using a modified version of the Gillespie algorithm, that can simulate multiple time delayed reactions (chemical reactions where each of the products is provided a time delay that determines when will it be released in the system as a "finished product").

For example, basic transcription of a gene can be represented by the following single-step reaction (RNAP is the RNA polymerase, RBS is the RNA ribosome binding site, and Pro$_i$ is the promoter region of gene i):

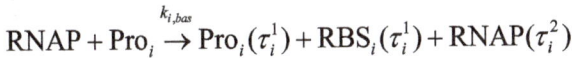

$$\text{RNAP} + \text{Pro}_i \xrightarrow{k_{i,bas}} \text{Pro}_i(\tau_i^1) + \text{RBS}_i(\tau_i^1) + \text{RNAP}(\tau_i^2)$$

Furthermore, there seems to be a trade-off between the noise in gene expression, the speed with which genes can switch, and the metabolic cost associated their functioning. More specifically, for any given level of metabolic cost, there is an optimal trade-off between noise and processing speed and increasing the metabolic cost leads to better speed-noise trade-offs.

A recent work proposed a simulator (SGNSim, *Stochastic Gene Networks Simulator*), that can model GRNs where transcription and translation are modeled as multiple time delayed events and its dynamics is driven by a stochastic simulation algorithm (SSA) able to deal with multiple time delayed events. The time delays can be drawn from several distributions and the reaction rates from complex functions or from physical parameters. SGNSim can generate ensembles of GRNs within a set of user-defined parameters, such as topology. It can also be used to model specific GRNs and systems of chemical reactions. Genetic perturbations such as gene deletions, gene over-expression, insertions, frame shift mutations can also be modeled as well.

The GRN is created from a graph with the desired topology, imposing in-degree and out-degree distributions. Gene promoter activities are affected by other genes expression products that act as inputs, in the form of monomers or combined into multimers and set as direct or indirect. Next, each direct input is assigned to an operator site and different transcription factors can be allowed, or not, to compete for the same operator site, while indirect inputs are given a target. Finally, a function is assigned to each gene, defining the gene's response to a combination of transcription factors (promoter state). The transfer functions (that is, how genes respond to a combination of inputs) can be assigned to each combination of promoter states as desired.

In other recent work, multiscale models of gene regulatory networks have been developed that focus on synthetic biology applications. Simulations have been used that model all biomolecular interactions in transcription, translation, regulation, and induction of gene regulatory networks, guiding the design of synthetic systems.

Prediction

Other work has focused on predicting the gene expression levels in a gene regulatory network. The approaches used to model gene regulatory networks have been constrained to be interpretable and, as a result, are generally simplified versions of the network. For example, Boolean networks have been used due to their simplicity and ability to handle noisy data but lose data information by having a binary representation of the genes. Also, artificial neural networks omit using a hidden layer so that they can be interpreted, losing the ability to model higher order correlations in the data. Using a model that is not constrained to be interpretable, a more accurate model can be produced. Being able to predict gene expressions more accurately provides a way to explore how drugs affect a system of genes as well as for finding which genes are interrelated in a process. This has been encouraged by the DREAM competition which promotes a competition for the best prediction algorithms. Some other recent work has used artificial neural networks with a hidden layer.

Gene Regulation at the Single Cell Level

Gene regulation is an intricate complex process, which involves genes, mRNAs and proteins that dictate cellular phenotypes and their response to external stimuli. Recent approaches employing genomics and proteomics and interactomic studies have helped probe the structure and signalling of these complex networks. However, more interesting aspects of the cell systems can be explored through the dynamics executed by these networks. To probe the dynamics of gene networks is complex because of the following reasons:

i. Simultaneous time resolved measurements of network components are required for dynamics studies and they generally are not accurate.

ii. There is variability in responses even among the cells that are genetically identical. Variation in cell parameters like cell size, stage of the cell cycle, metabolite concentration and intrinsic stochasticity of biochemical reactions inside the cell compel the cells to a differential response to the same external stimuli. Such variations in dynamics are often measured by distributions of relevant observations in a population which can be measured by techniques such as flow cytometry.

iii. Avoiding ambiguity in interpreting dynamic data requires tracking the fates of individual cells in the population for a time span greater than the process time. Since this time span equals several cell division cycles, collection and interpreting data over cell division cycles is a predominant challenge.

Visualization of gene expression and single cell measurements at the intracellular and intercellular levels at high resolution have made possible to understand the dynamics of gene regulation in these circuits. Imaging mRNA activity using FISH, MS2-GFP, molecular beacons, FRET and fluorescent microscopy at single molecule resolution have given us a handle to understand the gene expression at single cell level. The stochastic nature of gene expression and regulation are well studied using engineered gene networks.

Newer and evolving techniques in the field have enabled real time measurement of RNA. A major difficulty in using single fused fluorescence proteins is that they sometimes fluoresce more than the cellular auto fluorescence. The MS2 tagging system and RNA reporter RNA plasmid helps address this question. Once both the proteins are expressed in live cells the multiple fluorophores fuse to MS2 capsid proteins, bind to the MS2 binding sites in the UTR of RNA of interest. This technique gives a strong fluorescent signal, which allows visualization of an individual RNA molecule. Real time tracking of target RNA's can also be done using in vivo hybridization with molecular beacons. Molecular beacons are single stranded nucleic acid probes which contain a fluorophore and a quencher that woo apart upon binding to target RNA sequence.

Interpreting Protein Dynamics

The green fluorescence protein (GFP) and its derivative proteins facilitate real time visualization of proteins. Multi coloured fluorescence microscopy enables simultaneous measurement of multiple protein concentrations and the relative roles of intrinsic and extrinsic noise in gene expression. Fluorescence Resonance Energy Transfer (FRET) utilizes a donor and acceptor fluorophore to visualize conformational changes in individual molecules. The donor fluorophore transfers energy to the acceptor when the two come together and change the wavelength of the fluorescence signal.

Fluorescence based techniques in imaging provide the flexibility to study gene expression in single cells. Such studies show fluctuations in identical cells leading to the thought that gene regulation predominantly controls cellular properties. The stochastic variation in gene expression induces switching response in single cells creating binary 'ON', 'OFF' switches. Literature on signal processing in single cells (Issacs et al., Science. 2005), documents the relation between input and output signals in engineered E.coli networks.

The relationship between the gene expression rate and the abundance of regulatory proteins in single cells has been modelled using GRF, a gene regulation function (Rosenfeld et al., Science. 2005). This study involved an engineered two-step regulatory cascade in which the gene regulatory function in individual cells where measured dynamically and simultaneously with the input output signals. Time-lapse microscopy is employed to measure the GRF with population averaging. This interesting experiment observed dynamic fluctuations of GRF in individual cells implying that stochasticity in gene expression and minor variations in parameters limit the signal transfer property of transcription networks.

Noise propagation through a three-step transcription regulatory cascade is probed by the Van Oudenaarden group in E.coli systems. These experiments investigated the abundance of gene products as different steps in the cascade and attempted to correlate these evidences in single cell. This is done by measuring the expression changes in the input and output genes by varying the concentration of the inducers in the circuit.

Gene expression variations in individual cells may also influence the physiological states with the endogenous pathways (Acar et al., Nature. 2005). The galactose regulatory pathway in Saccharomyces cerevesiae was used as a model system to investigate the determinants of stability in cellular memory. Constructed signalling networks in E. coli, have been shown to generate bistability. Such networks have also been shown to store memory through the creation of discrete states.

Spatio Temporal Analysis of Gene Expression using Single Molecule Techniques

While fluorophore tags enable detection of target molecules and their localization in vivo fluorescence microscopy is widely adapted to determine the mean fluorescence of the tagged proteins and also to locate single molecules. The high diffusion rate of single molecules inside live cells makes their imaging difficult. This problem can be overcome by reducing their diffusion rate by localizing the molecules to the membrane where they diffuse slower than in the cytoplasm. A focused laser beam in a confocal microscope achieves single molecule detection near field scanning optical microscopy (NSOM), stochastic optical reconstruction microscopy, photo activated localization microscopy and stimulated emission and depression (STED) are the most powerful techniques for sub diffraction microscopy. These require scanning of large areas with a small window and hence are too slow for characterizing live cell dynamics. In order to explain the network dynamics completely, a mechanistic understanding of the networks is important. This approach requires high Spatio temporal resolution of fluorescent imaging and simultaneous measurement of the expression of network components because study of the average properties of cell cultures is not enough and more important is the sources of fluctuation and inter cellular variability. Micro fluidics and optogenetics help track multiple species of RNA and proteins in vivo with sub cellular evolution. Mechanical stimuli can regulate gene expression in live cells. The response of an individual cell modification to a defined molecular environment depends on the cell type and the mechanical stimulus. The mechanical stimuli require to elicit gene expression in live cells have been probed using AFM, optical tweezers and magnetic beads.

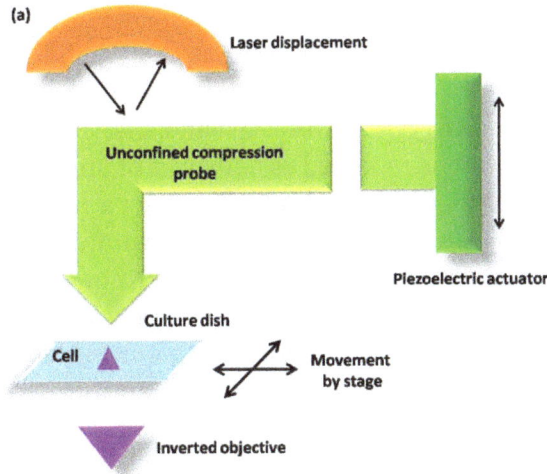

(a)

Laser displacement

Unconfined compression probe

Piezoelectric actuator

Culture dish

Cell

Movement by stage

Inverted objective

This device exposes single adherent cells to unconfined creep compression at forces ranging from 10nN to 200nN or grater. Bean theory was used to calculate the force while measuring the resulting cellular deformation

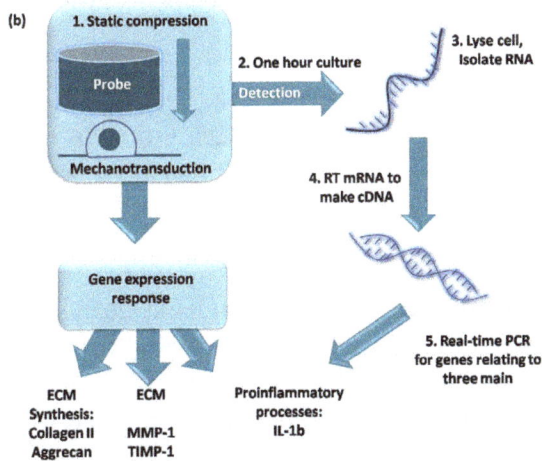

(b)

1. Static compression

Probe

2. One hour culture

Detection

3. Lyse cell, Isolate RNA

Mechanotransduction

4. RT mRNA to make cDNA

Gene expression response

5. Real-time PCR for genes relating to three main

ECM Synthesis: Collagen II Aggrecan

ECM MMP-1 TIMP-1

Proinflammatory processes: IL-1b

Unconfined compressions of single cells modify cell gene expression by mechanotransduction. For chondrocytes, this modification could occur in genes related to major processes involved in cartilage homeostasis and disease. Gene expression was measured in compressed chondrocytes through single-cell RT-PCR

Biological Network

A biological network is any network that applies to biological systems. A network is any system with sub-units that are linked into a whole, such as species units linked into a whole food web. Biological networks provide a mathematical representation of connections found in ecological, evolutionary, and physiological studies, such as neural networks. The analysis of biological networks with respect to human diseases has led to the field of network medicine.

Network Biology and Bioinformatics

Complex biological systems may be represented and analyzed as computable networks. For example, ecosystems can be modeled as networks of interacting species or a protein can be modeled as a network of amino acids. Breaking a protein down farther, amino acids can be represented as a network of connected atoms, such as carbon, nitrogen, and oxygen. Nodes and edges are the basic components of a network. Nodes represent units in the network, while edges represent the interactions between the units. Nodes can represent a wide-array of biological units, from individual organisms to individual neurons in the brain. Two important properties of a network are degree and betweenness centrality. Degree (or connectivity, a distinct usage from that used in graph theory) is the number of edges that connect a node, while betweenness is a measure of how central a node is in a network. Nodes with high betweenness essentially serve as bridges between different portions of the network (i.e. interactions must pass through this node to reach other portions of the network). In social networks, nodes with high degree or high betweenness may play important roles in the overall composition of a network.

As early as the 1980s, researchers started viewing DNA or genomes as the dynamic storage of a language system with precise computable finite states represented as a finite state machine. Recent complex systems research has also suggested some far-reaching commonality in the organization of information in problems from biology, computer science, and physics, such as the Bose–Einstein condensate (a special state of matter).

Bioinformatics has increasingly shifted its focus from individual genes, proteins, and search algorithms to large-scale networks often denoted as -omes such as biome, interactome, genome and proteome. Such theoretical studies have revealed that biological networks share many features with other networks such as the Internet or social networks, e.g. their network topology.

Networks in Biology

Protein–Protein Interaction Networks

Many protein–protein interactions (PPIs) in a cell form *protein interaction networks* (PINs) where proteins are *nodes* and their interactions are *edges*. PINs are the most intensely analyzed networks in biology. There are dozens of PPI detection methods to identify such interactions. The yeast two-hybrid system is a commonly used experimental technique for the study of binary interactions.

Recent studies have indicated conservation of molecular networks through deep evolutionary time. Moreover, it has been discovered that proteins with high degrees of connectedness are more likely to be essential for survival than proteins with lesser degrees. This suggests that the overall composition of the network (not simply interactions between protein pairs) is important for the overall functioning of an organism.

Gene Regulatory Networks (DNA–Protein Interaction Networks)

The activity of genes is regulated by transcription factors, proteins that typically bind to DNA. Most transcription factors bind to multiple binding sites in a genome. As a result, all cells have complex gene regulatory networks. For instance, the human genome encodes on the order of 1,400 DNA-binding transcription factors that regulate the expression of more than 20,000 human genes. Technologies to study gene regulatory networks include ChIP-chip, ChIP-seq, CliP-seq, and others.

Gene Co-expression Networks (Transcript–Transcript Association Networks)

Gene co-expression networks can be interpreted as association networks between variables that measure transcript abundances. These networks have been used to provide a systems biologic analysis of DNA microarray data, RNA-seq data, miRNA data etc. weighted gene co-expression network analysis is widely used to identify co-expression modules and intramodular hub genes. Co-expression modules may correspond to cell types or pathways. Highly connected intramodular hubs can be interpreted as representatives of their respective module.

Metabolic Networks

The chemical compounds of a living cell are connected by biochemical reactions which convert one compound into another. The reactions are catalyzed by enzymes. Thus, all compounds in a cell are parts of an intricate biochemical network of reactions which is called metabolic network. It is possible to use network analyses to infer how selection acts on metabolic pathways.

Signaling Networks

Signals are transduced within cells or in between cells and thus form complex signaling networks. For instance, in the MAPK/ERK pathway is transduced from the cell surface to the cell nucleus by a series of protein–protein interactions, phosphorylation reactions, and other events. Signaling networks typically integrate protein–protein interaction networks, gene regulatory networks, and metabolic networks.

Neuronal Networks

The complex interactions in the brain make it a perfect candidate to apply network theory. Neurons in the brain are deeply connected with one another and this results in complex networks being present in the structural and functional aspects of the brain. For instance, small-world network properties have been demonstrated in connections between cortical areas of the primate brain. This suggests that cortical areas of the brain are not directly interacting with each other, but most areas can be reached from all others through only a few interactions.

Food Webs

All organisms are connected to each other through feeding interactions. That is, if a species eats or is eaten by another species, they are connected in an intricate food web of predator and prey interactions. The stability of these interactions has been a long-standing question in ecology. That is to say, if certain individuals are removed, what happens to the network (i.e. does it collapse or adapt)? Network analysis can be used to explore food web stability and determine if certain network properties result in more stable networks. Moreover, network analysis can be used to determine how selective removals of species will influence the food web as a whole. This is especially important considering the potential species loss due to global climate change.

Between-species Interaction Networks

In biology, pairwise interactions have historically been the focus of intense study. With the recent advances in network science, it has become possible to scale up pairwise interactions to include individuals of many species involved in many sets of interactions to understand the structure and function of larger ecological networks. The use of network analysis can allow for both the discovery and understanding how these complex interactions link together within the system's network, a property which has previously been overlooked. This powerful tool allows for the study of various types of interactions (from competitive to cooperative) using the same general framework. For example, plant-pollinator interactions are mutually beneficial and often involve many different species of pollinators as well as many different species of plants. These interactions are critical to plant reproduction and thus the accumulation of resources at the base of the food chain for primary consumers, yet these interaction networks are threatened by anthropogenic change. The use of network analysis can illuminate how pollination networks work and may in turn inform conservation efforts. Within pollination networks, nestedness (i.e., specialists interact with a subset of species that generalists interact with), redundancy (i.e., most plants are pollinated by many pollinators), and modularity play a large role in network stability. These network properties may actually work to slow the spread of disturbance effects through the system and potentially buffer the pollination network from anthropogenic changes somewhat. More generally, the structure of species interactions within an ecological network can tell us something about the diversity, richness, and robustness of the network. Researchers can even compare current constructions of species interactions networks with historical reconstructions of ancient networks to determine how networks have changed over time. Recent research into these complex species interactions networks is highly concerned with understanding what factors (e.g., diversity) lead to network stability.

Within-species Interaction Networks

Network analysis provides the ability to quantify associations between individuals, which makes it possible to infer details about the network as a whole at the species and/or population level. Researchers interested in animal behavior across a multitude of taxa, from insects to primates, are starting to incorporate network analysis into their research. Researchers interested in social insects (e.g., ants and bees) have used network analyses to better understand division of labor, task allocation, and foraging optimization within colonies; Other researchers are interested in how certain network properties at the group and/or population level can explain individual level behaviors. For instance, a study on wire-tailed manakins (a small passerine bird) found that a male's degree in the network largely predicted the ability of the male to rise in the social hierarchy (i.e. eventually obtain a territory and matings). In bottlenose dolphin groups, an individual's degree and betweenness centrality values may predict whether or not that individual will exhibit certain behaviors, like the use of side flopping and upside-down lobtailing to lead group traveling efforts; individuals with high betweenness values are more connected and can obtain more information, and thus are better suited to lead group travel and therefore tend to exhibit these signaling behaviors more than other group members. Network analysis can also be used to describe the social organization within a species more generally, which frequently reveals important proximate mechanisms promoting the use of certain behavioral strategies. These descriptions are frequently linked to ecological properties (e.g., resource distribution). For example, network analyses revealed subtle differences in the group dynamics of two related equid fission-fusion species, Grevy's zebra and onagers, living in variable environments; Grevy's zebras show distinct preferences in their association choices when they fission into smaller groups, whereas onagers do not. Similarly, researchers interested in primates have also utilized network analyses to compare social organizations across the diverse primate order, suggesting that using network measures (such as centrality, assortativity, modularity, and betweenness) may be useful in terms of explaining the types of social behaviors we see within certain groups and not others. Finally, social network analysis can also reveal important fluctuations in animal behaviors across changing environments. For example, network analyses in female chacma baboons (*Papio hamadryas ursinus*) revealed important dynamic changes across seasons which were previously unknown; instead of creating stable, long-lasting social bonds with friends, baboons were found to exhibit more variable relationships which were dependent on short-term contingencies related to group level dynamics as well as environmental variability. This is a very small set of broad examples of how researchers can use network analysis to study animal behavior. Research in this area is currently expanding very rapidly. Social network analysis is a valuable tool for studying animal behavior across all animal species, and has the potential to uncover new information about animal behavior and social ecology that was previously poorly understood.

Network Motifs

Let us try to decipher the complex networking inside cells now. Every living cell is comprised of intricate networks that represent interactions and inter dependent events among the bio molecules in the network. Gene regulation networks, protein- protein interaction networks, metabolic pathways, and neural networks are well known examples of this intricacy.

The primary aim of bio medical research in the post genomic era has been to catalogue all the bio molecules and their interactions within a living cell. Advances in the field of network biology such as high throughput data collection techniques, micro arrays, protein chips, automated yeast two hybrid strains permit simultaneous investigation of cellular components and information on the molecular interaction between these components.

These networks are not independent but instead form a 'network of networks' which regulate cellular behavior. The intricate complexity of the network poses a major challenge to contemporary biology and demands an integrated theoretical and experimental approach that can map out and model the topological and dynamic properties of these networks.

This class will focus on tools which can quantify network parameters and therefore help us understand the cell's internal organization and evolution. I am sure this will completely alter our perspective of a cell and the cellular mechanisms in place. Understanding the organising principles of cellular network leads to a better understanding of the cell as a system and also provides relevance for experimental biologists in the role of individual molecules in various cellular processes.

Cellular Networks-Nomenclatures

All of us know that complex systems ranging from the internet to a living cell behave due to the orchestrated activity of network components that show pair wise and more complex interactions. These components can be reduced to a series of 'nodes' connected to each other by certain 'links', each link representing the interaction between two components. These nodes and links constitute a network. This is also called 'a graph' in the mathematical language. Physical interactions such as protein- protein interactions, protein-nucleic acid interactions and protein metabolites interactions can be concretely explained with the node link nomenclature.

Networks can be classified as 'directed' or 'undirected' based on the nature of the interaction. Figure illustrates this through an elegant graphical representation.

Directed Network

In such networks as shown in Fig (a) the interaction between any two nodes has a well defined direction like the direction of signalling from a transcription factor to a gene or the direction of material flow from a substrate to a product in a metabolic reaction.

Undirected Network

These are networks as shown in Fig (b) in which the links have not been assigned a specific direction.eg. In protein interact networks a link represents a mutual binding relationship. If protein A binds to protein B then protein B also binds to protein A.

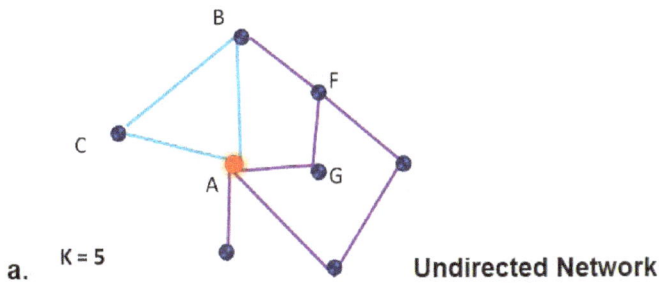

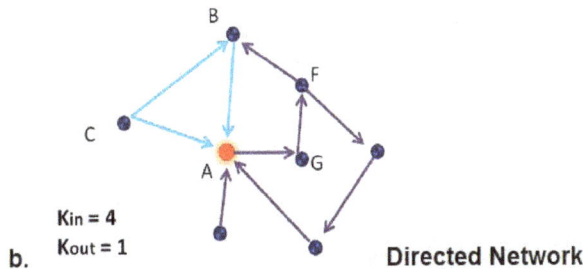

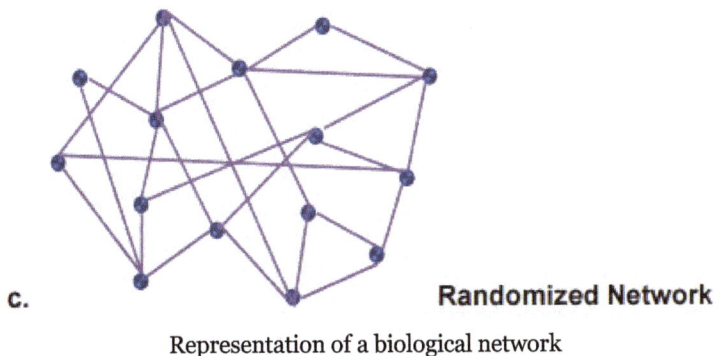

a. $K = 5$ **Undirected Network**

b. $K_{in} = 4$, $K_{out} = 1$ **Directed Network**

c.   **Randomized Network**

Representation of a biological network

Patterns, Randomized Networks and Network Motifs in Biology

Network motifs are simple building blocks of complex networks and are statistically over represented sub structures or sub graphs in a network. Since the number of sub graphs in biological networks increases exponentially with the network or motif size, it is difficult to detect larger network motifs in a biological network. Hence network motifs can be defined as recurring patterns of interactions that are significantly over represented in a biological system. This over representation of the sub network indicates the functional importance of such motif. Therefore it becomes important to explore these abundant motifs in biological networks. Milo et al, (Science 2002) in one of their break through explorations ,work analyzed 18 different networks from

1. Transcription networks of E.coli and S.cerevisiae

2. Synaptic connections between neurons in c. elegans

3. Throphic interactions between predator and prey in ecological systems

As represented in Fig (a) in each of these networks the number of nodes is represented by 'n'. In each of the above mentioned cases all possible motifs of size n=3 and n=4 were enumerated and compared to an average count over thousand random networks. Randomized networks were generated without compromising on certain properties of the original network.

1. In-degree, out- degree and mutual degree: this is done by swapping edges to generate random graphs.

2. The number of appearances of all n-1 node sub graphs for n>3. This is done to ensure that high significance is assigned to highly significant sub pattern also.

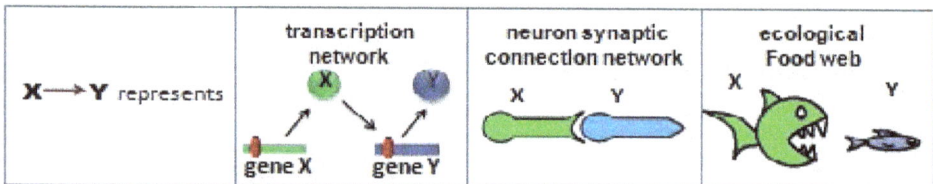

(a) Types of network investigated in Milo et. al (Science 2002)

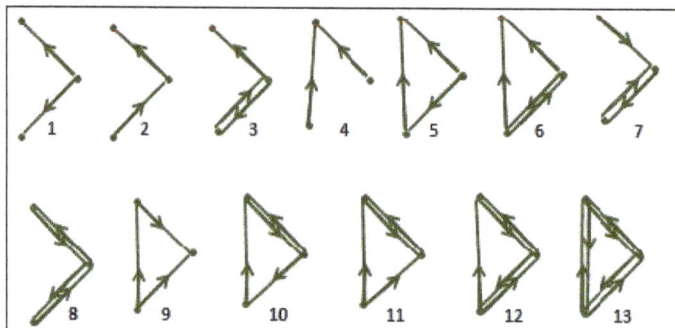

(b) Directed motifs for n=3 where n refers to the number of nodes.

Identification of Network Motifs

The following rules help us identify network motifs in biological systems.

1. Identification all sub graphs of n nodes in the network

2. Randomization of the network without changing the number of nodes, edges and degree distribution.

3. Identification of all sub graphs of n nodes in the randomized network.

4. Comparison of more frequently occurring significant sub graphs with randomized ones in the network and designating them as motifs.

Real live transcription networks of organisms like E.coli show numerous patterns of nodes and edges but it is important to look for meaningful patterns that are statistically significant to derive biological information that brings us to discuss the concept of randomized networks.

Randomized networks are a type of networks that posses the same characteristics of a real network. These have the same numbers of nodes and edges as in biological systems. But in these networks random connections are made between nodes and edges. Network motifs are patterns that occur more significantly and more often in real networks than in randomized networks. Recurrent patterns give us the basic idea that these must have been evolutionarily conserved against mutation.

The best way to illustrate this is the fact that edges are easily lost in a transcription network and a single mutation abolishes transcription factor binding hence facilitating the loss of edge in the network. In the same way mutations which generate a binding site for transcription factor X in promoter region of gene Y can help add new edges to the network. Mutations, duplication events that reposition pieces on the genome or insertion events can generate new binding sites and hence add new edges to the network. This clearly demonstrates that the occurrence of the network motifs more often than in randomized network proves that this selection offers an advantage to the organism.

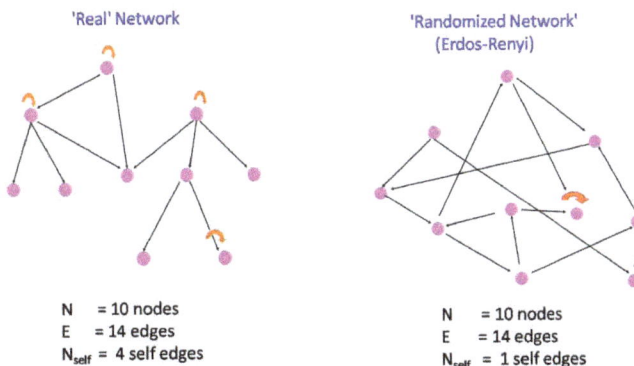

'Real' Network

'Randomized Network'
(Erdos-Renyi)

N = 10 nodes	N = 10 nodes
E = 14 edges	E = 14 edges
N_{self} = 4 self edges	N_{self} = 1 self edges

Representation of Real and Randomized Network with nodes, edges and self edges

Auto Regulation

Self Edges

Self edges are those edges that originate and end at the same node. The *E.coli* network has approximately 40 self edges each of which correspond to transcription factors that regulate the transcription of their own genes. This type of regulation of a gene by its own gene product is called autogenous control or auto regulation. 34 of the auto regulatory proteins in the *E.coli* network have been found to repress their own transcription. This process is referred to as negative auto regulation as referred to in Figure.

Negative auto regulation is a network motif and occurs at higher numbers than expected in random networks. Such structures display engineering advantages and are more prone to evolutionary selection in their appearances network motif.

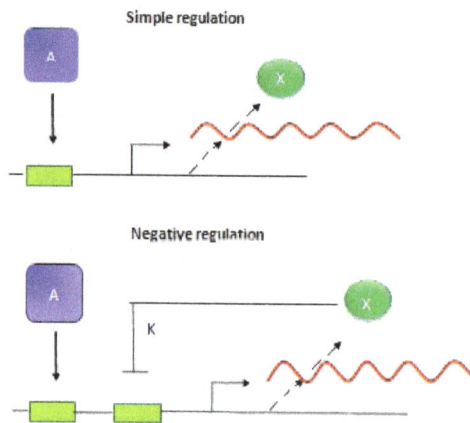

Schematic representation of Simple and Negative Auto Regulation in Biological Networks

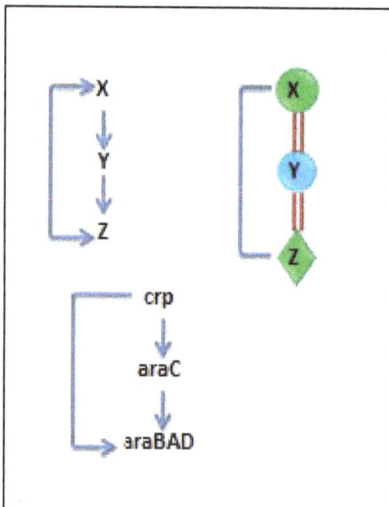

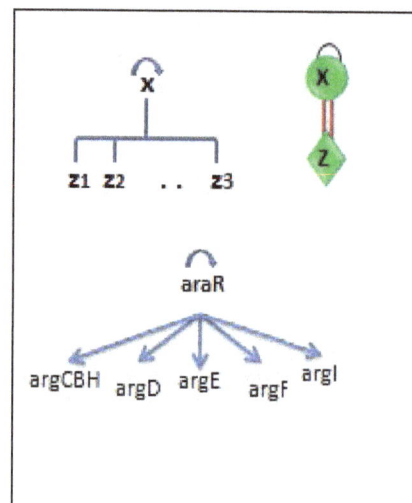

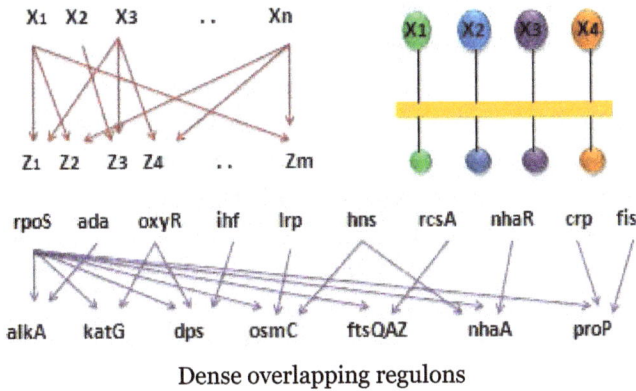

Dense overlapping regulons

Feed Forward Loops

In order to understand the general features of such networks and to extract useful information from them, we dissect them at hierarchical levels- into modules and motifs which can explain their functionality, evolution and dynamic behavior. Over the process of evolution, these networks show information processing functions. Interesting investigations on network behaviour have shown that simple switching circuits, amplifiers or oscillators can map to the core process of biological decision making. These have been implemented by two or three gene network motifs and are characterized by how they behave around fixed points in the system. Here the steady state of the system as well as the process of achieving equilibrium in the system reflects the characteristic function performed by the genetic circuit.

Network motifs appear at frequencies much higher than those expected at random and hence imply information processing roles for these motifs. To arrive at such significant patterns, one first identifies the different patterns of these motifs in real and randomized networks and then calculates the number of appearances of these patterns in the real and random networks. The discussion that follows focuses on patterns with 3 nodes (forming a triangle), Figure below. There are 13 possible 3-node patterns in such arrangement. Of these, only one of them qualifies to be a network motif called the Feed Forward Loop.

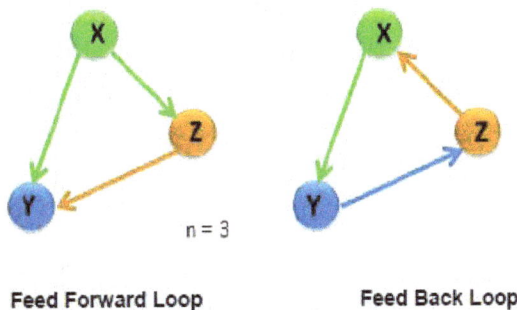

Feed Forward Loop   Feed Back Loop

Representative Feed Forward and Feed Back Loops with nodes n=3 forming a triangle

The most significant of the network motifs in E. coli and yeast is the Feed Forward Loop which is defined by a transcription factor X that regulates a second transcription factor Y. X and Y both jointly regulate an operon Z by binding to its regulatory region. Here X is called the general transcription factor, Y the specific transcription factor and Z the effector operon. As described in Figure, this type of motif occurs in the L-arabinose utilization system where Crp is the general transcription factor and Ara-C is a specific transcription factor. Such a motif characterizes 40 effector operons in 22 different systems in the network database and accommodates 10 different transcription factors.

Feed Forward Loop

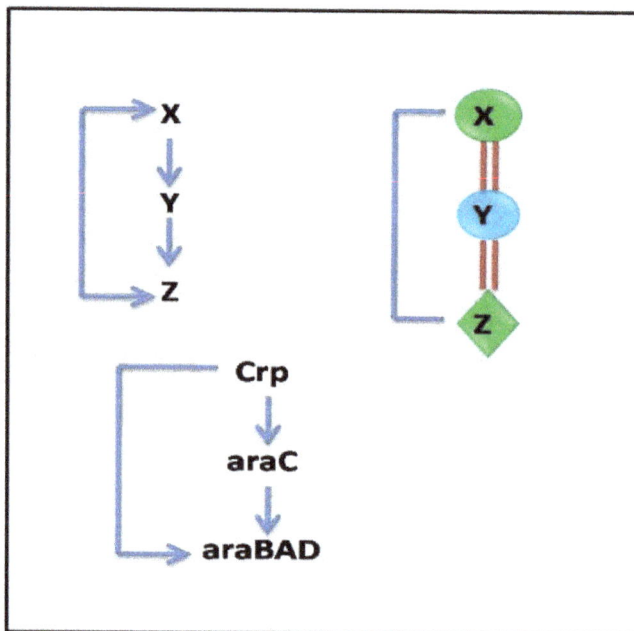

Diagrammatic representation of FFL with an example of L-arabinose utilization system

In order to appreciate the dynamics of the Feed Forward Loop, one needs to study the regulation at each one of its three edges, each of which can represent activation or a repressive interaction. S_X and S_Y are the two input signals to the Feed Forward Loop. The signals could be small molecules, protein partners, biochemical or environmental stimuli or covalent modifications that activate or inhibit the transcription of X and Y. Therefore there are 8 possible types of FFL - 8 structural configurations of activator and repressor interactions.

Coherent and Incoherent Feed Forward Loops

A Feed Forward Loop is termed 'coherent' if the direct effect of the general transcription factor X on the effector operons has the same sign (positive or negative) as its net indirect effect through the specific transcription factor.

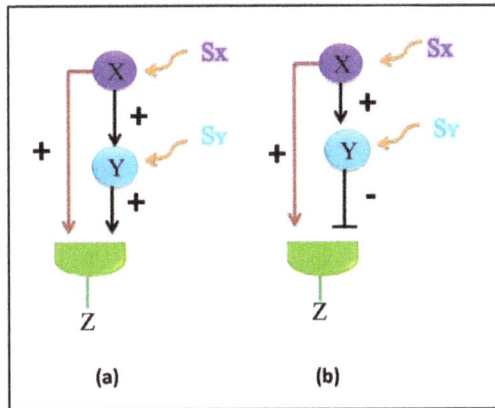

Feed-Forward Loop (FFL) (a) Coherent FFL (type1), X regulates Y and both jointly regulate Z.
(b) Incoherent FFL (type1), here Sx and Sy are the signals

In other words, if X regulates Y positively and if X and Y both positively regulate Z, the Feed Forward Loop is coherent. i.e. the sign of the direct path of regulation (X to Z) is the same as the overall sign of the indirect regulation path (X to Z through Y). The overall sign of a path is determined by the multiplication of the sign of each arrow on the path. In Figure (a) we see that the sign of the indirect path (X→Y→Z) is plus x plus =plus, while the direct path (X→Z) is already plus. Since both the direct and indirect paths have the same positive sign, this loop is called a Coherent Feed Forward Loop.

The other type of FFL is called Incoherent FFL in which the sign of the indirect path of regulation is opposite to that of the direct path. In type-1 Incoherent FFL as denoted the direct path is positive and the indirect path is negative. The Incoherent FFLs show odd number of minus edges.

In both the coherent and incoherent loops, the effects of the general and specific transcription factors X and Y are integrated at the promoter region of gene Z. The expression profile of Z is modulated by the concentrations of X and Y bound to their inducers. The cis regulatory input function of Z describes this modulation. cis regulatory input functions include logic gates like AND which require both X and Y to express Z and OR gates in which either X or Y is sufficient to express Z.

Both Coherent and Incoherent Feed Forward Loops are sign sensitive. Type 1 coherent FFLs (in which all three regulations are positive) are the most abundant type of Feed

The Incoherent Feed Forward loop type-1 is the second most abundant type of FFL among biological networks. The other types of feedforward loop do not appear more frequently than CFFL I and ICFFL I.

Coherent Feed Forward Loops

The Feed Forward Loop is a network motif constituted by a transcription factor X which regulates another transcription factor Y and both X and Y regulate a gene

Z. Such Feed Forward Loops comprise two parallel paths that regulate the circuit, a direct path from X->Z and an indirect path of X->Z through Y. A Coherent Feed Forward Loop has the same overall sign for the indirect path as that of the direct path of regulation. Though all the Feed Forward Loops appear with equal frequency across transcription networks, Type -1 Coherent FFL is the most abundant type of FFL and shows a positive sign for all the regulations involved in the circuit.

In order to explain the functional logic exhibited by the Feed Forward Loop, let us consider a typical situation inside a cell during the process of transcription. Let the cell express multiple copies of a protein X, one of the transcription factors in the Feed Forward Loop. Let signal S_x be the input to X. This X is inactive without the signal S_*. At a time t=0, the strong signal S_x triggers the activation of X, transiting it to a form X^*. This process is called a step like stimulation of X.

X^*, the active protein now binds to the promoter of gene Y initiates production of protein Y, the second transcription factor in the Feed Forward Loop. Mean while X^* binds to the promoter of gene Z. If the circuit follows an AND logic (input function at the Z promoter), X^* alone cannot activate the production of Z. it requires the binding of both X^* and Y^*. This indicates that Y should build up to sufficient levels of expression to cross the activation threshold of gene Z (K_{YZ}).Activation of Z also requires the presence of S_y which activates Y to its form Y^*. Therefore the appearance of the signal S_x and Y are both required to activate Z. This introduces a delay in the production of Z. This function is represented by the truth tables below.

Coherent Feed Forward Loop Type I (CFFL I)

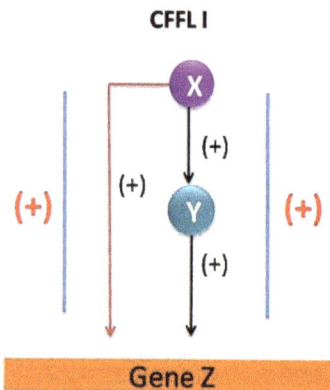

Z Logic	AND	OR
Steady- state		
Z(Sx, Sy)	Sx ∧ Sy	Sx
Response delay		
Sx ON step	Delay	--
Sx OFF step	--	Delay
Inverted out	No	No

(a) Coherent Feed Forward Loop

Table (a) AND- and OR- gates at the Z Type I (CFFL I) promoter for CFFL I

Case 1: Steady State

In an AND logic circuit, described in Figure (a) when both the signals S_x and S_y are present. In the presence of signals the system reaches a steady state. The active part of transcription factor X which is X* binds to the promoter gene Z through gene Y and X binds to Z independent of Y. So gene Z receives two inputs (one from gene X and the other from Y) leading to the expression of both the genes.

In short, X controls Z, forming a direct pathway. X controls Z, through Y forming an indirect pathway.

Case 2: S_X ON State

Let us now consider the case when S_x is ON. In the AND circuit, When SX is on, the transcription factor of the protein X becomes active X*and binds to the promoter of gene Z, thus expressing Z. (direct path).

The activated protein X* also binds to promoter of the gene Y and activates Y (Y*) as a result of which gene Y gets expressed. Since Y should reach the activation threshold, to bind to the promoter of gene Z (indirect path), the process requires time. Hence delaying the expression of Z.

Case 3: S_x OFF State

In the AND circuit, when S_x is OFF, it cannot trigger the expression of Z directlu or through Y. Hence there is no expression of gene Z.

In an OR function, when S_x is OFF, the transcription factor of the protein X does not become active X* and does not bind to the promoter of gene Z and gene Z is not expressed(direct path). The protein X does not binds to promoter of the gene Y and gene Y is not fully expressed. When Y reaches the activation threshold, it binds to the promoter of gene Z (indirect path). Since the levels of expression of Y isles in this case, the process is delayed and so is the expression of gene Z.

Case 4: Inverted Out

In both AND circuit, when the system is inverted, this case results in no expression of Z. Similarly OR gate can be worked out for all the four cases.

Coherent Feed Forward Loop type II (CFFL II)

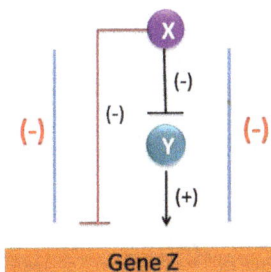

	CFFL II	
Z Logic	AND	OR
Steady- state		
$Z(S_X, S_Y)$	$\hat{S}_X \wedge S_Y$	$\hat{S}_X$
Response delay		
Sx ON step	--	Delay
Sx OFF step	Delay	--
Inverted out	Yes	Yes

(b) Coherent Feed Forward Loop

Table (b) AND- and OR- gates at the Z Type II (CFFL II) promoter for CFFL II

Case 1: Steady state

In the AND circuit, the system reaches a steady state when $\hat{Sx}$ (the inverted input signal of S_x) and signal S_y are present. In the presence of $\hat{Sx}$ signal, the active part of the transcription factor X which is X^* binds to the promoter of gene Z through gene Y and also binds to gene Z independent of gene Y. Hence both the genes are expressed in this case.

In an OR circuit, the system requires only the inverted signal Sx ($\hat{Sx}$) to reach steady state.

Case 2: S_X ON State

In an AND circuit, when S_x is ON, there is no gene expression.

The activated protein X* also binds to promoter of the gene Y and represses gene Y. But Y reaches the activation threshold K and binds to the promoter of gene Z (indirect path). Since this process requires time for Y to reach the threshold. The expression of gene Z is delayed.

Case 3: S_X OFF State

In the AND circuit, when S_x is OFF, the transcription factor of the protein X does not get activated X* and hence does not bind to the promoter of gene Z. Hence is not repressed (direct path).

The transcription factor of the protein X does not bind to promoter of the gene Y and gene Y is also not repressed. Y reaches the activation threshold, it binds to the promoter of gene Z (indirect path). Since this process depends on the activation of Y and the time required to reach the threshold. The expression of gene Z is delayed.

Case 4: Inverted Out

In both AND circuit, when the system is inverted, there is expression of gene Z without any delay, as its gene expression is controlled by the active X*(direct path).

Similarly OR circuit can be worked out for all the four cases.

Coherent Feed Forward Loop type III (CFFL III)

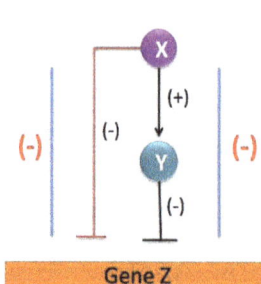

(c) Coherent Feed Forward Loop

CFFL III

Z Logic	AND	OR
Steady- state		
$Z(Sx, Sy)$	$\hat{Sx}$	$\hat{Sx} \wedge \hat{Sy}$
Response delay		
Sx ON step	--	--
Sx OFF step	Delay	Delay
Inverted out	Yes	Yes

Table (c) AND- and OR- gates at the Z Type III (CFFL III) promoter for CFFL III

Case 1: Steady State:

In the AND circuit, the system reaches steady state when the signals inverted S_x is present. In the presence of inverted Sx signal, the active part of the transcription factor of X which is X^* binds to the promoter of gene Z.

Case 2: S_X ON State

In an AND, when Sx is ON, there is no gene expression of Z.

Case 3: S_X OFF State

In an AND circuit, when S_X is OFF, the transcription factor of the protein X cannot make an active X*and hence does not bind to the promoter of gene Z and gene Z is not repressed (direct path).

The transcription factor of the protein X does not bind to promoter of the gene Y and gene Y is also not expressed. Hence gene Y will not repress gene Z because there is no production of Y (indirect path). The expression of gene Z is delayed.

Case 4: Inverted Out

In both AND circuit and OR circuit, when the system is inverted, there is expression of gene Z with delay, as its gene expression is controlled only by the active X* (direct path).

Similarly OR circuit can be worked out for all the four cases.

Coherent Feed Forward Loop Type IV (CFFL IV)

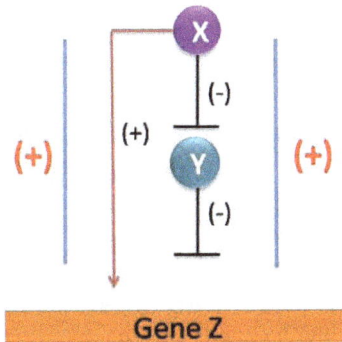

(d) Coherent Feed Forward Loop

CFFL IV		
Z Logic	**AND**	**OR**
Steady-state		
$Z(S_X, S_Y)$	S_X	$S_X \vee \check{S}_Y$
Response delay		
Sx ON step	Delay	Delay
Sx OFF step	--	--
Inverted out	No	No

Table (d) AND- and OR- gates at the Z Type IV (CFFL IV) promoter for CFFL IV

Case 1: Steady State

In an AND circuit, the system reaches steady state when the signal S_X is present. In the presence of the signal S_X, the active part of the transcription factor of X which is X^* binds to the promoter of both gene Z through gene Y and also to gene Z independent of gene Y. Therefore gene Z is expressed even when gene Y is repressed. The repression of gene Z does not take place through gene Y.

Case 2: S_X ON State

In both AND circuits, when S_X is on, the transcription factor of the protein X becomes active X*and binds to the promoter of gene Z and expresses gene Z (direct path).The activated protein X* also binds to promoter of the gene Y and represses gene Y. The repression of gene Z does not take place through Y. So the expression of gene Z is delayed.

Case 3: S_X OFF State

In both AND circuit, when S_X is OFF, there is no gene expression.

Case 4: Inverted Out

In both AND circuit, when the system is inverted, there is no expression of gene Z. Similarly OR circuit can be worked out all the four cases.

Besides the logic functions and truth tables discussed elaborately in this lesson, Coherent Feed Forward Loops have well interesting properties in biological systems. An outstanding example is the function of Type-1 AND gate Coherent FFL as a sign sensitive delay in the arabinose system in E.coli in the presence of cyclic AMP signals. We shall deal with these in the next few classes which bring out a clear understanding of the information processing roles of the biological networks.

Kinetic Response of Feed Forward Loops

The kinetic response of the Feed Forward Loops are interesting to observe in living cells because they regulate functional responses in the circuit through decision logic. We shall now try to understand how a Coherent Feed Forward Loop can be used to induce a sign dependent delay in the circuit and how such sign sensitive delays are relevant to the arabinose system of *E.coli*. We shall also discuss how an Incoherent FFL accelerates the response time in the circuit, tracing the example of the galactose system of *E.coli*.

In the earlier class, we learnt how the Coherent FFL Type-1 with AND logic introduces a delay when the signal S_X is ON. Let us recall that in the Type-1 Coherent FFL (AND gate), when S_X takes a step addition, the activator Y grows to a sufficient concentration and crosses the activation threshold. The expression of Z begins only now. Therefore there is a delay introduced in the circuit. This behavior is called sign sensitive delay in the circuit because the delay is dependent on whether the step is ON or OFF. The speed of the response of the circuit is characterized by the term 'response time', defined as the time taken by the gene product Z to reach 50% of its physiologically determined steady state level of expression. Let us consider a circuit where the pulse S_X (ON pulse) appears for brief time.

When S_X is suddenly removed following an OFF step, S_X becomes inactive, unbinds from the promoters Y and Z. therefore the AND gate requires only one of its inputs to go OFF in order to stop the expression of Z. Thus an OFF step causes no delay in the dynamics of Z.

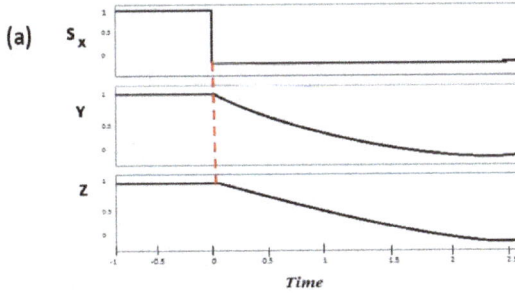

The C1-FFL AND gate doesnot introduce a delay during an OFF step of S_X
The Fig depicts the dynamics of the FFL during an OFF step of SX at time t=0

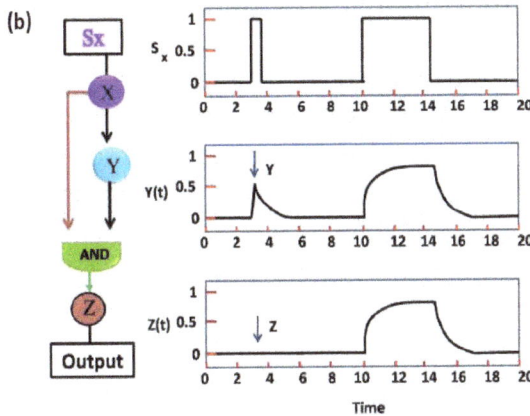

The C1-FFL AND gate behaves as persistence detector. On receiving a signal S_X, Y takes time to accumulate and cross the activation threshold to activate Z. Since the persistent pulse causes a delay in the production of Z, the circuit is called a persistent detector circuit.

Persistent Detector

If the input ON pulse is shorter than the delay time TON, the input will not result in the expression of Z in the C-1 FFL circuit. This is because the gene Y cannot cross its activation threshold during the brief period of the ON pulse. This indicates that only pulse is longer than T_{ON} lead to the expression of Z. Since such persistent pulses lead to the expression of Z, this Feed Forward Loop C-1 FFL is called a 'persistent detector' as shown in Figure for ON pulses. The same circuit responds immediately to OFF pulses. The major difference in the behavior of C-1 FFL to a simple regulatory circuit which is not an FFL is that a simple regulatory circuit does not filter out short input pulses and shows expression of Z as long as the input pulse is present.

OR gate C-1 FFL Functions as a Sign Sensitive Delay Circuit During OFF Steps of S_X

Let us now replace the AND gate in the C-1 FFL with an OR gate at the Z promoter. How does the response of the circuit change? When the C-1 FFL has an OR gate, the gene Z gets activated immediately when S_X is ON. This is because one ON input is enough to activate the function of an OR gate. Hence no delay is introduced following the ON step of S_X. But Z is inactive due to a delay following an OFF step because both the inputs go OFF for the OR gate to be inactivated. Thus Y^* can activate Z even without X^*. It takes time for Y^* to decay after an OFF step of S. Therefore the Coherent

FFL Type-1 with an OR function also behaves as a sign sensitive delay element with signs opposite to that of its AND version. Experimental measurements on gene expression at high resolution have observed such a dynamic circuit which induce a delay with S_X OFF during OR function in the flagella system of *E.coli*. The OR gate C-1 FFL has been shown to regulate the production of proteins that self assemble into a motor which facilitates *E.coli* swimming through rotational movement of the flagella. Accurate measurements under experimental conditions have shown that the delay after removal of the input S_X is around one hour (one cell generation time). The delay has the same order of magnitude as the time it takes to assemble a flagella motor. The interesting feature with the OR gate C-1 FFL is that it continues to express the gene for about an hour even after the input signal fades. This means that the OR C-1 FFL is capable of protecting the genetic circuit against transient loss of input signal.

Significance of Sign Sensitive Delay in Biological Networks

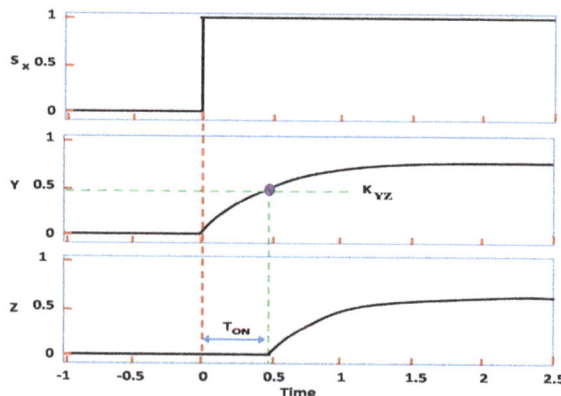

The c1-FFL OR gate is a Sign Senstive Delay circuit when S_X is OFF

As discussed, sign sensitive delays protect circuits against brief fluctuations in input. Evolutionary demands might have positioned Type-1 CFFLs in diverse systems inside a cell which requires protection against transient loses. We know that the cellular environment is continuous and highly fluctuating system where the stimulating signals could prevail for very brief periods which may not be sufficient to elicit a response. Since the sign sensitive delay element in C-1 FFL also acts as an asymmetric filter it confers an advantageous filtering function to the fluctuating environment inside a cell. This leads us to believe that the natural selection of a Feed Forward Loop is primarily based on the filtering function we can provide to the cells.

Sign Sensitive Delay in Arabinose System

The proteins in the arabinose system transport the sugar arabinose to the cell and break it down to form an energy or carbon source. Generally, the cell prefers glucose to arabinose. Hence, the sugars arabinose and glucose are the inputs for an arabinose system to make the decision for uptake. Obviously, the prerequisite for the proteins to be formed inside the arabinose system is the presence of arabinose, and not glucose. The absence of glucose in a cell comes to light by the formation of a small molecule named cAMP. The two transcription activators, namely CRP (detects cAMP) and araC (detects arabinose) are connected in a C1-FFL with in an AND input function, as shown in Figure. Experiments in the environment of *E. coli* revealed that the delay in the arabinose FFL is on the same order of magnitude as the duration of spurious pulses of the input signal that are generally observed when *E.coli* transits between different growth conditions.

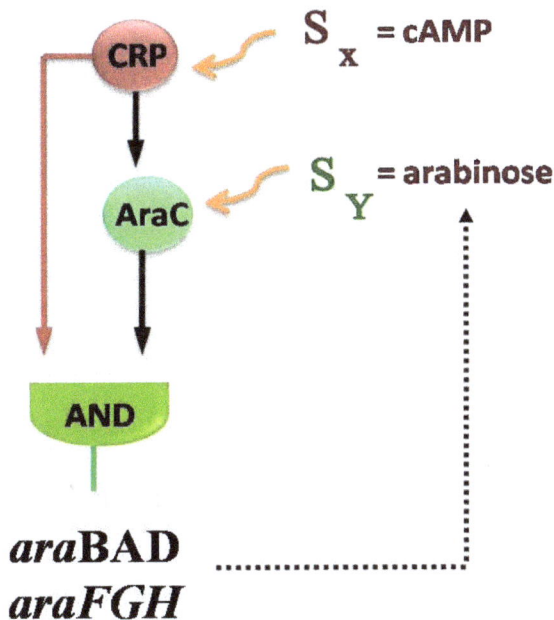

FFL in the arabinose system of E. coli

Thus, the FFL in the system, apart from being able to filter out the short fluctuations in the input signal, responds only to persistent stimuli for instance, glucose starvation, which need to make use of sugar arabinose. Though embedded in additional interactions, the FFL in arabinose system shows sign-sensitive delay. In spite of the three-gene FFL circuit in isolation, the arabinose FFL shows the expected dynamics also when embedded within the interaction networks of the cell.

Transcription Networks

The living cell is a complex machinery involving thousands of interacting proteins and other biological molecules. These cells utilise proteins based on the situations they encounter. During sugar sensing, the cells begin to secrete proteins that transport sugar into the cells. Cells respond to damage by producing damage repair proteins. Every cell thus senses its environment continuously, regulates protein production and maintains cellular homeostasis. Live cells as we know are kept dynamic through gene expression programs that involve the transcription of thousands of genes in a coordinated manner. As we all know, the expression of a gene is facilitated by transcription regulatory proteins which recognize specific promoter sequences. The association of regulatory proteins with genes across a genome and the continuous cascade of information processing which senses the rate of production of a particular protein inside a cell constitute a transcription regulatory network. These are in general called transcription networks. Metabolic networks describe the possible pathways that a cell may use to accomplish metabolic processes. In a similar way the map of the transcriptional regulatory network of an organism describes potential pathways the cells of the organisms utilise to regulate global gene expression programs. This network map establishes a high connectivity between gene expression programs and cellular functions through networks of transcriptional regulatory molecules which in turn regulate other molecular players in transcription.

Transcription Network-Elements

We know that transcription networks depict the interaction between transcription factors and genes. The process of transcription ensures that RNA polymerase produces mRNA (messenger RNA) corresponding to the coding sequence of a gene. The mRNA is translated to form a protein which is also referred to as the gene product. Each gene is preceded by a small stretch of regulatory DNA called promoter. The promoter is a specific DNA sequence that can bind RNA polymerase, which is a complex of several proteins. The rate at which the gene is transcribed is determined by the quality of the binding site and the promoter

which controls the number of mRNA produced per unit time. The RNA polymerase complex acts on a number of genes while the transcription factors regulate changes in expression profiles of specific genes. The transcription factors when bound change the probability per unit time of RNA polymerase binding to the promoter to produce the mRNA. Transcription factors can also function as activators which increase the transcription rate of a gene or as repressors which reduce the transcription rate. The transcription factor proteins themselves are encoded by genes which are regulated by other transcription factors. These transcription factors might have been regulated by other set of transcription factors in the genome.

Such a complexity leads to a set of interactions which may be visualised to form a transcription network. Thus a transcription network depicts all interactions of the transcription regulatory proteins inside the cell. In such a network the genes are represented by nodes and the edges represents the transcription regulation of one gene by the protein product of the other. Therefore in a transcription network the directed edge X➔Y means that the product of a gene X is a transcription factor protein which controls the rate of transcription of gene Y (by binding to the promoter of gene Y) , which is explained in the following Figure.

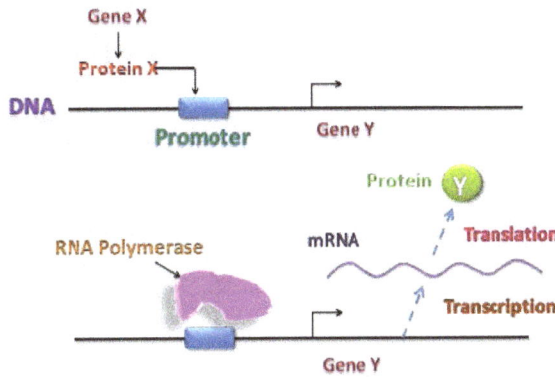

A schematic of transcription and translation of a gene Y

As shown in the Figure, the network has inputs called signals which represent information from the environment. A signal could be a protein modification, environmental signal, biochemical stimuli, small molecule or a molecular partner that influences one of the transcription factors in the network. The signals initiate a physical change in the transcription factor protein and switch it to an active molecular state. The signal S_X therefore shifts the gene X to its active state X*, binds to the promoter of the gene Y, increasing the rate of transcription and hence increasing the production of protein Y. The gene network is thus a dynamic system. With the arrival of the input signal S_X gene activation profiles change, activities of transcription factors change resulting in changes in protein production rate.

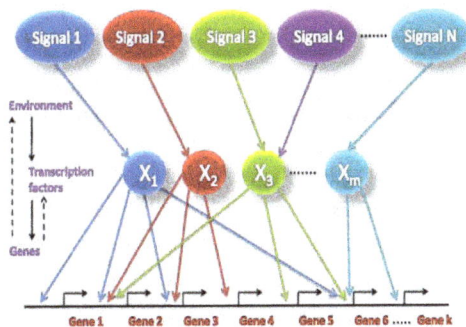

A typical Transcription Network depicting the interaction between the Signals, Transcription Factors and Genes

Timescales of Transcription Networks

The timescales at which reactions happen in transcription network are quite interesting to note. Table explain timescales of reactions taking place in *E. coli*. We can observe that the input signals such as biochemical stimuli or temperature change the activity of transcription factors on sub second (milliseconds) timescale. The binding of the active transcription factor to its DNA site attains equilibrium in seconds. The transcription and translation of the target gene happens in minutes while the accumulation if protein products and change in the concentration of the stable protein takes hours.

Table: Timescales of reactions in the Transcription Network of the Bacterium *E. coli*	
Binding of a small molecule (a signal) to a transcription factor, causing a change in transcription factor activity	~1 ms
Binding of active transcription factor to its DNA site	~1 sec
Transcription + Translation of the gene	~5 min
Timescale for 50% change in concentration of the translated protein (stable proteins)	~1 h (one cell generation)

Modularity of Transcription Networks

A remarkable property of transcription networks is the modularity of the components of the network. This allows taking off the DNA of a gene from one organism and expressing it on the other. The best example is the isolation of the green fluorescent protein (GFP) from the genome of a jelly fish and to introduce this gene into bacteria causing the bacteria to express green fluorescence.

This modularity confers plasticity to transcription networks during evolution, facilitating introduction of newer genes and newer regulatory pathways. The edges in transcription networks appear to evolve on a faster timescale than the coding regions of the genes. For example, mice and humans are closely related and have similar genes. But their transcription regulation mechanisms and timescale of protein production are quite different.

Signs on the Edges

Each edge in a transcription network corresponds to an interaction in which a transcription factor directly controls the transcription rate of a gene. Such interactions can be classified into two types.

1.   Activation or positive control

2.   Repression or negative control

Activation or Positive Control

When a transcription factor binds to the promoter and increases the rate of transcription it is referred to as activation or positive control. Figure clearly explains the concept further.

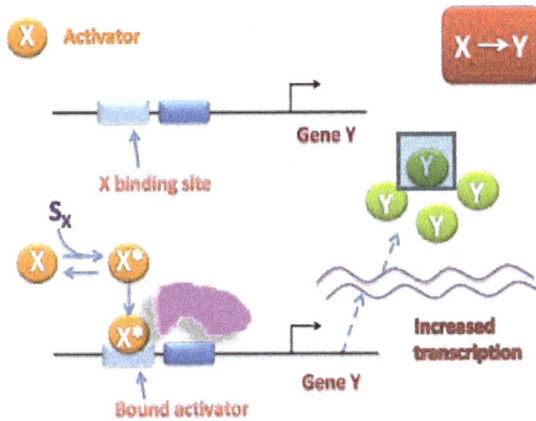

Binding of an activator protein (X*) iIncreases the rate of Transcription of gene Y

Repression or Negative Control

When a transcription factor binds to the promoter and decreases the rate of transcription it is referred to as repression or negative control. Figure further explains in detail.

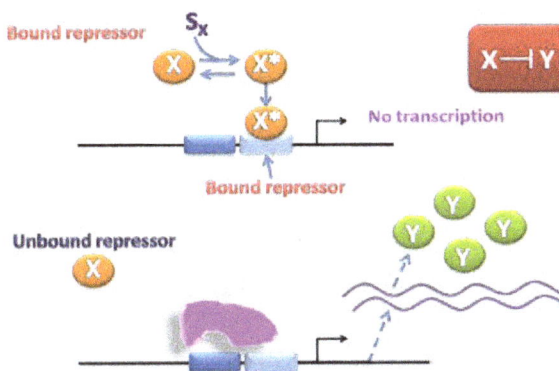

Bound and unbound repressor protein (X*) regulate the rate of transcription of gene Y.

Hence each edge in the network is represented by a plus sign for activation and a minus sign for repression. Generally transcription networks show more positive interactions than negative interactions. For example organisms such as *E. coli* and yeast show 60-80% activation interactions

Biological Network Inference

Biological network inference is the process of making inferences and predictions about biological networks.

Biological Networks

In a topological sense, a network is a set of nodes and a set of directed or undirected edges between the nodes. Many types of biological networks exist, including transcriptional, signalling and metabolic. Few such networks are known in anything approaching their complete structure, even in the simplest bacteria. Still less is known on the parameters governing the behavior of such networks over time, how the networks at different levels in a cell interact, and how to predict the complete state description of a eukaryotic cell or bacterial organism at a given point in the future. Systems biology, in this sense, is still in its infancy.

There is great interest in network medicine for the modelling biological systems. This article focuses on a necessary prerequisite to dynamic modeling of a network: inference of the topology, that is, prediction of the "wiring diagram" of the network. More specifically, we focus here on inference of biological network structure using the growing sets of high-throughput expression data for genes, proteins, and metabolites. Briefly, methods using high-throughput data for inference of regulatory networks rely on searching for patterns of partial correlation or conditional probabilities that indicate causal influence. Such patterns of partial correlations found in the high-throughput data, possibly combined with other supplemental data on the genes or proteins in the proposed networks, or combined with other information on the organism, form the basis upon which such algorithms work. Such algorithms can be of use in inferring the topology of any network where the change in state of one node can affect the state of other nodes.

Transcriptional Regulatory Networks

Genes are the nodes and the edges are directed. A gene serves as the source of a direct regulatory edge to a target gene by producing an RNA or protein molecule that functions as a transcriptional activator or inhibitor of the target gene. If the gene is an activator, then it is the source of a positive regulatory connection; if an inhibitor, then it is the source of a negative regulatory connection. Computational algorithms take as primary input data measurements of mRNA expression levels of the genes under consideration for inclusion in the network, returning an estimate of the network topology. Such algorithms are typically based on linearity, independence or normality assumptions, which must be verified on a case-by-case basis. Clustering or some form of statistical classification is typically employed to perform an initial organization of the high-throughput mRNA expression values derived from microarray experiments, in particular to select sets of genes as candidates for network nodes. The question then arises: how can the clustering or classification results be connected to the underlying biology? Such results can be useful for pattern classification – for example, to classify subtypes of cancer, or to predict differential responses to a drug (pharmacogenomics). But to understand the relationships between the genes, that is, to more precisely define the influence of each gene on the others, the scientist typically attempts to reconstruct the transcriptional regulatory network. This can be done by data integration in dynamic models supported by background literature, or information in public databases, combined with the clustering results. The modelling can be done by a Boolean network, by Ordinary differential equations or Linear regression models, e.g. Least-angle regression, by Bayesian network or based on Information theory approaches. For instance it can be done by the application of a correlation-based inference algorithm, as will be discussed below, an approach which is having increased success as the size of the available microarray sets keeps increasing

Signal Transduction

Signal transduction networks (very important in the biology of cancer). Proteins are the nodes and directed edges represent interaction in which the biochemical conformation of the child is modified by the action of the parent (e.g. mediated by phosphorylation, ubiquitylation, methylation, etc.). Primary input into the inference algorithm would be data from a set of experiments measuring protein activation / inactivation (e.g., phosphorylation / dephosphorylation) across a set of proteins. Inference for such signalling networks is complicated by the fact that total concentrations of signalling proteins will fluctuate over time due to transcriptional and translational regulation. Such variation can lead to statistical confounding. Accordingly, more sophisticated statistical techniques must be applied to analyse such datasets.

Metabolic

Metabolite networks. Metabolites are the nodes and the edges are directed. Primary input into an algorithm would be data from a set of experiments measuring metabolite levels.

Protein-protein Interaction

Protein-protein interaction networks are also under very active study. However, reconstruction of these networks does not use correlation-based inference in the sense discussed for the networks already described (interaction does not necessarily imply a change in protein state), and a description of such interaction network reconstruction.

Gene co-expression Network

A gene co-expression network (GCN) is an undirected graph, where each node corresponds to a gene, and a pair of nodes is connected with an edge if there is a significant co-expression relationship between them. Having gene expression profiles of a number of genes for several samples or experimental conditions, a gene co-expression network can be constructed by looking for pairs of genes which show a similar expression pattern across samples, since the transcript levels of two co-expressed genes rise and fall together across samples. Gene co-expression networks are of biological interest since co-expressed genes are controlled by the same transcriptional regulatory program, functionally related, or members of the same pathway or protein complex.

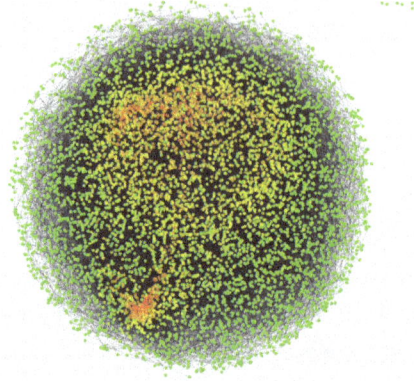

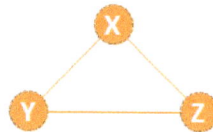

A gene co-expression network constructed from a microarray dataset containing gene expression profiles of 7221 genes for 18 gastric cancer patients

The direction and type of co-expression relationships are not determined in gene co-expression networks; whereas in a gene regulatory network (GRN) a directed edge connects two genes, representing a biochemical process such as a reaction, transformation, interaction, activation or inhibition. Compared to a GRN, a GCN does not attempt to infer the causality relationships between genes and in a GCN the edges represent only a correlation or dependency relationship among genes. Modules or the highly connected subgraphs in gene co-expression networks correspond to clusters of genes that have a similar function or involve in a common biological process which causes many interactions among themselves.

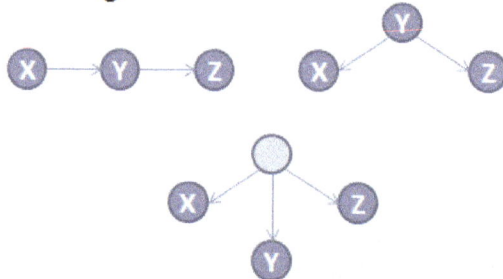

The direction of edges is overlooked in gene co-expression networks. While three genes X, Y and Z are found to be co-expressed, it is not determined whether X activates Y and Y activates Z, or Y activates X and Z, or another gene activates three of them.

Gene co-expression networks are usually constructed using datasets generated by high-throughput gene expression profiling technologies such as Microarray or RNA-Seq.

History

The concept of gene co-expression networks was first introduced by Butte and Kohane in 1999 as *relevance networks*. They gathered the measurement data of medical laboratory tests (e.g. hemoglobin level) for a number of patients and they calculated the Pearson correlation between the results for each pair of tests and the pairs of tests which showed a correlation higher than a certain level were connected in the network (e.g. insulin level with blood sugar). Bute and Kohane used this approach later with mutual information as the co-expression measure and using gene expression data for constructing the first gene co-expression network.

Constructing Gene Co-expression Networks

A good number of methods have been developed for constructing gene co-expression networks. In principle, they all follow a two step approach: calculating co-expression measure, and selecting significance threshold. In first step, a co-expression measure is selected and a similarity score is calculated for each pair of genes using this measure. Then, a threshold is determined and gene pairs which have a similarity score higher than the selected threshold are considered to have significant co-expression relationship and are connected by an edge in the network.

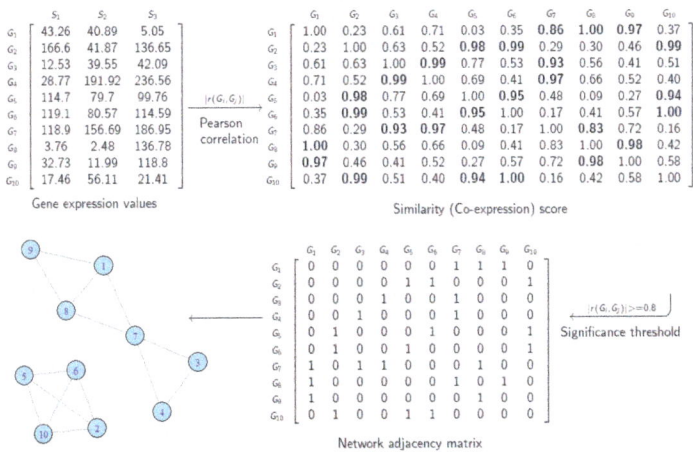

Gene expression values

	S_1	S_2	S_3
G_1	43.26	40.89	5.05
G_2	166.6	41.87	136.65
G_3	12.53	39.55	42.09
G_4	28.77	191.92	236.56
G_5	114.7	79.7	99.76
G_6	119.1	80.57	114.59
G_7	118.9	156.69	186.95
G_8	3.76	2.48	136.78
G_9	32.73	11.99	118.8
G_{10}	17.46	56.11	21.41

$|r(G_i, G_j)|$ Pearson correlation →

Similarity (Co-expression) score

	G_1	G_2	G_3	G_4	G_5	G_6	G_7	G_8	G_9	G_{10}
G_1	1.00	0.23	0.61	0.71	0.03	0.35	0.86	1.00	0.97	0.37
G_2	0.23	1.00	0.63	0.52	0.98	0.99	0.29	0.30	0.46	0.99
G_3	0.61	0.63	1.00	0.99	0.77	0.53	0.93	0.56	0.41	0.51
G_4	0.71	0.52	0.99	1.00	0.69	0.41	0.97	0.66	0.52	0.40
G_5	0.03	0.98	0.77	0.69	1.00	0.95	0.48	0.09	0.27	0.94
G_6	0.35	0.99	0.53	0.41	0.95	1.00	0.17	0.41	0.57	1.00
G_7	0.86	0.29	0.93	0.97	0.48	0.17	1.00	0.83	0.72	0.16
G_8	1.00	0.30	0.56	0.66	0.09	0.41	0.83	1.00	0.98	0.42
G_9	0.97	0.46	0.41	0.52	0.27	0.57	0.72	0.98	1.00	0.58
G_{10}	0.37	0.99	0.51	0.40	0.94	1.00	0.16	0.42	0.58	1.00

$|r(G_i, G_j)| >= 0.8$ Significance threshold

Network adjacency matrix

	G_1	G_2	G_3	G_4	G_5	G_6	G_7	G_8	G_9	G_{10}
G_1	0	0	0	0	0	0	1	1	1	0
G_2	0	0	0	0	1	1	0	0	0	1
G_3	0	0	0	1	0	0	1	0	0	0
G_4	0	0	1	0	0	0	1	0	0	0
G_5	0	1	0	0	0	1	0	0	0	1
G_6	0	1	0	0	1	0	0	0	0	1
G_7	1	0	1	1	0	0	0	1	0	0
G_8	1	0	0	0	0	0	1	0	1	0
G_9	1	0	0	0	0	0	0	1	0	0
G_{10}	0	1	0	0	1	1	0	0	0	0

The two general steps for constructing a gene co-expression network: calculating co-expression score (e.g. the absolute value of Pearson correlation coefficient) for each pair of genes, and selecting a significance threshold (e.g. correlation > 0.8).

The input data for constructing a gene co-expression network is often represented as a matrix. If we have the gene expression values of m genes for n samples (conditions), the input data would be an $m \times n$ matrix, called expression matrix. For instance, in a microarray experiment the expression values of thousands of genes are measured for several samples. In first step, a similarity score (co-expression measure) is calculated between each pair of rows in expression matrix. The resulting matrix would be an $m \times m$ matrix, called similarity matrix. Each element in this matrix shows how similar the expression level of two genes change together. In second step, the elements in similarity matrix which are above a certain threshold (i.e. significant co-expressions) are replaced by 1 and the remaining elements are replaced by 0. The resulting matrix, called adjacency matrix, represents the graph of the constructed gene co-expression network. In this matrix, each element shows whether two genes are connected in the network (the 1 elements) or not (the 0 elements).

Co-expression Measure

The expression values of a gene for different samples can be represented as a vector, thus calculating the co-expression measure between a pair of genes is the same as calculating the selected measure for two vectors of numbers.

Pearson's correlation coefficient, Mutual Information, Spearman's rank correlation coefficient and Euclidean distance are the four mostly used co-expression measures for constructing gene co-expression networks. Euclidean distance measures the geometric distance between two vectors, and so considers both the direction and the magnitude of the vectors of gene expression values. Mutual information measures how much knowing the expression levels of one gene reduces the uncertainty about the expression levels of another. Pearson's correlation coefficient measures the tendency of two vectors to increase or decrease together, giving a measure of their overall correspondence. Spearman's rank correlation is the Pearson's correlation calculated for the ranks of gene expression values in a gene expression vector. Several other measures such as partial correlation, regression, and combination of partial correlation and mutual information have also been used.

Each of these measures have their own advantages and disadvantages. The Euclidean distance is not appropriate when the absolute levels of functionally related genes are highly different. Furthermore, if two genes have consistently low expression levels but are otherwise randomly correlated, they might still appear close in Euclidean space. One advantage to mutual information is that it can detect non-linear relationships; however this can turn into a disadvantage because of detecting sophisticated non-linear relationships which does not look biologically meaningful. In addition, for calculating mutual information one should estimate the distribution of the data which needs a large number of samples for a good estimate. Spearman's rank correlation coefficient is more robust to outliers, but on the other hand it is less sensitive to expression values and in datasets with small number of samples may detect many false positives.

Pearson's correlation coefficient is the most popular co-expression measure used in constructing gene co-expression networks. The Pearson's correlation coefficient takes a value between -1 and 1 where absolute values close to 1 show strong correlation. The positive values correspond to an activation mechanism where the expression of one gene increases with the increase in the expression of its co-expressed gene and vice versa. When the expression value of one gene decreases with the increase in the expression of its co-expressed gene, it corresponds to an underlying suppression mechanism and would have a negative correlation.

There are two disadvantages for Pearson correlation measure: it can only detect linear relationships and it is sensitive to outliers. Moreover, Pearson correlation assumes that the gene expression data follow a normal distribution. Song et al. have suggested *biweight midcorrelation (bicor)* as a good alternative for Pearson's correlation. "Bicor is a median based correlation measure, and is more robust than the Pearson correlation but often more powerful than the Spearman's correlation". Furthermore, it has been shown that "most gene pairs satisfy linear or monotonic relationships" which indicates that "mutual information networks can safely be replaced by correlation networks when it comes to measuring co-expression relationships in stationary data".

Threshold Selection

Several methods have been used for selecting a threshold in constructing gene co-expression networks. A simple thresholding method is to choose a co-expression cutoff and select relationships which their co-expression exceeds this cutoff. Another approach is to use Fisher's Z-transformation which calculates a z-score for each correlation based on the number of samples. This z-score is then converted into a p-value for each correlation and a cutoff is set on the p-value. Some methods permute the data and calculate a z-score using the distribution of correlations found between genes in permuted dataset. Some other approaches have also been used such as threshold selection based on clustering coefficient or random matrix theory.

The problem with p-value based methods is that the final cutoff on the p-value is chosen based on statistical routines(e.g. a p-value of 0.01 or 0.05 is considered significant), not based on a biological insight.

WGCNA is a framework for constructing and analyzing weighted gene co-expression networks. The WGCNA method selects the threshold for constructing the network based on the scale-free topology of gene co-expression networks. This method constructs the network for several thresholds and selects the threshold which leads to a network with scale-free topology. Moreover, the WGCNA method constructs a weighted network which means all possible edges appear in the network, but each edge has a weight which shows how significant is the co-expression relationship corresponding to that edge.

References

- Sirri V, Urcuqui-Inchima S, Roussel P, Hernandez-Verdun D (January 2008). "Nucleolus: the fascinating nuclear body". Histochem. Cell Biol. 129 (1): 13–31. doi:10.1007/s00418-007-0359-6. PMC 2137947. PMID 18046571

- Mashaghi, A.; et al. (2004). "Investigation of a protein complex network". European Physical Journal. 41 (1): 113–121. doi:10.1140/epjb/e2004-00301-0

- Köhler A, Hurt E (October 2007). "Exporting RNA from the nucleus to the cytoplasm". Nat. Rev. Mol. Cell Biol. 8 (10): 761–73. doi:10.1038/nrm2255. PMID 17786152

- Habibi, Iman; Emamian, Effat S.; Abdi, Ali (2014-10-07). "Advanced Fault Diagnosis Methods in Molecular Networks". PLOS ONE. 9 (10): e108830. doi:10.1371/journal.pone.0108830. ISSN 1932-6203. PMC 4188586. PMID 25290670

- Alberts, Bruce; Alexander Johnson; Julian Lewis; Martin Raff; Keith Roberts; Peter Walters (2002). "The Shape and Structure of Proteins". Molecular Biology of the Cell; Fourth Edition. New York and London: Garland Science. ISBN 0-8153-3218-1

- Jambhekar A, Derisi JL (May 2007). "Cis-acting determinants of asymmetric, cytoplasmic RNA transport". RNA. 13 (5): 625–42. doi:10.1261/rna.262607. PMC 1852811. PMID 17449729

- Romanuk, T.; et al. (2010). "Maintenance of positive diversity-stability relations along a gradient of environmental stress". PLoS ONE. 5: e10378. doi:10.1371/journal.pone.0010378. PMC 2860506. PMID 20436913

- Linksvayer, T.; et al. (2012). "Developmental evolution in social insects: Regulatory networks from genes to societies". Journal of Experimental Zoology Part B: Molecular and Developmental Evolution. 318: 159–169. doi:10.1002/jez.b.22001

- Prudovsky I, Tarantini F, Landriscina M, et al. (April 2008). "Secretion Without Golgi". J. Cell. Biochem. 103 (5): 1327–43. doi:10.1002/jcb.21513. PMC 2613191. PMID 17786931

- Jeremy M. Berg, John L. Tymoczko, Lubert Stryer; Web content by Neil D. Clarke (2002). "3. Protein Structure and Function". Biochemistry. San Francisco: W. H. Freeman. ISBN 0-7167-4684-0

- Amaral PP, Dinger ME, Mercer TR, Mattick JS (March 2008). "The eukaryotic genome as an RNA machine". Science. 319 (5871): 1787–9. Bibcode:2008Sci...319.1787A. doi:10.1126/science.1155472. PMID 18369136

- Berk V, Cate JH (June 2007). "Insights into protein biosynthesis from structures of bacterial ribosomes". Curr. Opin. Struct. Biol. 17 (3): 302–9. doi:10.1016/j.sbi.2007.05.009. PMID 17574829

- Bolouri, Hamid; Bower, James M. (2001). Computational modeling of genetic and biochemical networks. Cambridge, Mass: MIT Press. ISBN 0-262-02481-0

- Kober L, Zehe C, Bode J (April 2013). "Optimized signal peptides for the development of high expressing CHO cell lines". Biotechnol. Bioeng. 110 (4): 1164–73. doi:10.1002/bit.24776. PMID 23124363

- Faith JJ, Hayete B, Thaden JT, Mogno I, Wierzbowski J, Cottarel G, Kasif S, Collins JJ, Gardner TS (January 2007). "Large-scale mapping and validation of Escherichia coli transcriptional regulation from a compendium of expression profiles". PLoS Biology. 5 (1): e8. PMC 1764438. PMID 17214507. doi:10.1371/journal.pbio.0050008

- Moreau P, Brandizzi F, Hanton S, et al. (2007). "The plant ER-Golgi interface: a highly structured and dynamic membrane complex". J. Exp. Bot. 58 (1): 49–64. doi:10.1093/jxb/erl135. PMID 16990376

Fundamentals of Gene Expression Regulation

Mechanisms that enable the production of genetic material are collectively called gene expression regulation. Gene regulation controller exist in a gene regulatory network. The chapter serves as a source to understand the major categories related to gene expression regulation.

Regulation of Gene Expression

Regulation of gene expression includes a wide range of mechanisms that are used by cells to increase or decrease the production of specific gene products (protein or RNA), and is informally termed *gene regulation*. Sophisticated programs of gene expression are widely observed in biology, for example to trigger developmental pathways, respond to environmental stimuli, or adapt to new food sources. Virtually any step of gene expression can be modulated, from transcriptional initiation, to RNA processing, and to the post-translational modification of a protein. Often, one gene regulator controls another, and so on, in a gene regulatory network.

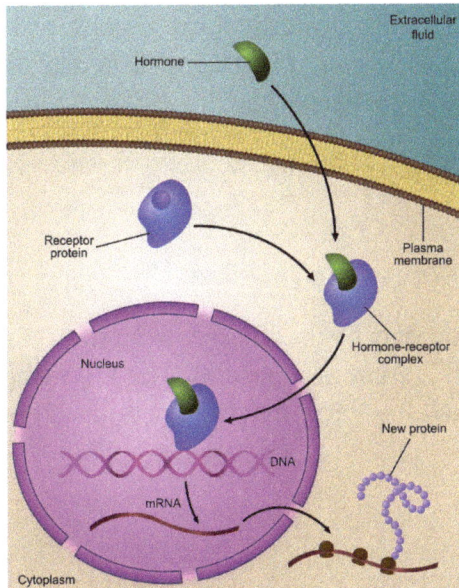

Regulation of gene expression by a hormone receptor

Diagram showing at which stages in the DNA-mRNA-protein pathway expression can be controlled

Gene regulation is essential for viruses, prokaryotes and eukaryotes as it increases the versatility and adaptability of an organism by allowing the cell to express protein when needed. Although as early as 1951, Barbara McClintock showed interaction between two genetic loci, Activator (*Ac*) and Dissociator (*Ds*), in the color formation of maize seeds, the first discovery of a gene regulation system is widely considered to be the identification in 1961 of the *lac* operon, discovered by Jacques Monod, in which some enzymes involved in lactose metabolism are expressed by *E. coli* only in the presence of lactose and absence of glucose.

In multicellular organisms, gene regulation drives cellular differentiation and morphogenesis in the embryo, leading to the creation of different cell types that possess different gene expression profiles from the same genome sequence. This explains how evolution actually works at a molecular level, and is central to the science of evolutionary developmental biology ("evo-devo").

The initiating event leading to a change in gene expression includes activation or deactivation of receptors.

Regulated Stages of Gene Expression

Any step of gene expression may be modulated, from the DNA-RNA transcription step to post-translational modification of a protein. The following is a list of stages where gene expression is regulated, the most extensively utilised point is Transcription Initiation:

- Chromatin domains
- Transcription

- Post-transcriptional modification

- RNA transport

- Translation

- mRNA degradation

Modification of DNA

In eukaryotes, the accessibility of large regions of DNA can depend on its chromatin structure, which can be altered as a result of histone modifications directed by DNA methylation, ncRNA, or DNA-binding protein. Hence these modifications may up or down regulate the expression of a gene. Some of these modifications that regulate gene expression are inheritable and are referred to as epigenetic regulation.

Structural

Transcription of DNA is dictated by its structure. In general, the density of its packing is indicative of the frequency of transcription. Octameric protein complexes called nucleosomes are responsible for the amount of supercoiling of DNA, and these complexes can be temporarily modified by processes such as phosphorylation or more permanently modified by processes such as methylation. Such modifications are considered to be responsible for more or less permanent changes in gene expression levels.

Chemical

Methylation of DNA is a common method of gene silencing. DNA is typically methylated by methyltransferase enzymes on cytosine nucleotides in a CpG dinucleotide sequence (also called "CpG islands" when densely clustered). Analysis of the pattern of methylation in a given region of DNA (which can be a promoter) can be achieved through a method called bisulfite mapping. Methylated cytosine residues are unchanged by the treatment, whereas unmethylated ones are changed to uracil. The differences are analyzed by DNA sequencing or by methods developed to quantify SNPs, such as Pyrosequencing (Biotage) or MassArray (Sequenom), measuring the relative amounts of C/T at the CG dinucleotide. Abnormal methylation patterns are thought to be involved in oncogenesis.

Histone acetylation is also an important process in transcription. Histone acetyltransferase enzymes (HATs) such as CREB-binding protein also dissociate the DNA from the histone complex, allowing transcription to proceed. Often, DNA methylation and histone deacetylation work together in gene silencing. The combination of the two seems to be a signal for DNA to be packed more densely, lowering gene expression.

Regulation of Transcription

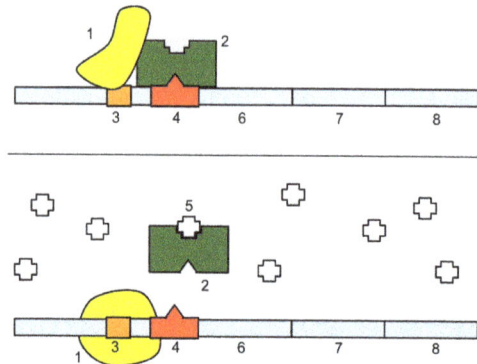

1: RNA Polymerase, *2*: Repressor, *3*: Promoter, *4*: Operator, *5*: Lactose, *6*: lacZ, *7*: lacY, *8*: lacA. Top: The gene is essentially turned off. There is no lactose to inhibit the repressor, so the repressor binds to the operator, which obstructs the RNA polymerase from binding to the promoter and making lactase. Bottom: The gene is turned on. Lactose is inhibiting the repressor, allowing the RNA polymerase to bind with the promoter, and express the genes, which synthesize lactase. Eventually, the lactase will digest all of the lactose, until there is none to bind to the repressor. The repressor will then bind to the operator, stopping the manufacture of lactase.

Regulation of transcription thus controls when transcription occurs and how much RNA is created. Transcription of a gene by RNA polymerase can be regulated by several mechanisms. Specificity factors alter the specificity of RNA polymerase for a given promoter or set of promoters, making it more or less likely to bind to them (i.e., sigma factors used in prokaryotic transcription). Repressors bind to the Operator, coding sequences on the DNA strand that are close to or overlapping the promoter region, impeding RNA polymerase's progress along the strand, thus impeding the expression of the gene. The image to the right demonstrates regulation by a repressor in the lac operon. General transcription factors position RNA polymerase at the start of a protein-coding sequence and then release the polymerase to transcribe the mRNA. Activators enhance the interaction between RNA polymerase and a particular promoter, encouraging the expression of the gene. Activators do this by increasing the attraction of RNA polymerase for the promoter, through interactions with subunits of the RNA polymerase or indirectly by changing the structure of the DNA. Enhancers are sites on the DNA helix that are bound by activators in order to loop the DNA bringing a specific promoter to the initiation complex. Enhancers are much more common in eukaryotes than prokaryotes, where only a few examples exist (to date). Silencers are regions of DNA sequences that, when bound by particular transcription factors, can silence expression of the gene.

Regulation of Transcription in Cancer

In vertebrates, the majority of gene promoters contain a CpG island with numerous CpG sites. When many of a gene's promoter CpG sites are methylated the gene becomes silenced. Colorectal cancers typically have 3 to 6 driver mutations and 33 to 66 hitchhiker or passenger mutations. However, transcriptional silencing may be of more importance than mutation in causing progression to cancer. For example, in colorectal cancers about 600 to 800 genes are transcriptionally silenced by CpG island methylation. Transcriptional repression in cancer can also occur by other epigenetic mechanisms, such as altered expression of microRNAs. In breast cancer, transcriptional repression of BRCA1 may occur more frequently by over-expressed microRNA-182 than by hypermethylation of the BRCA1 promoter.

Post-transcriptional Regulation

After the DNA is transcribed and mRNA is formed, there must be some sort of regulation on how much the mRNA is translated into proteins. Cells do this by modulating the capping, splicing, addition of a Poly(A) Tail, the sequence-specific nuclear export rates, and, in several contexts, sequestration of the RNA transcript. These processes occur in eukaryotes but not in prokaryotes. This modulation is a result of a protein or transcript that, in turn, is regulated and may have an affinity for certain sequences.

Three Prime Untranslated Regions and MicroRNAs

Three prime untranslated regions (3'-UTRs) of messenger RNAs (mRNAs) often contain regulatory sequences that post-transcriptionally influence gene expression. Such 3'-UTRs often contain both binding sites for microRNAs (miRNAs) as well as for regulatory proteins. By binding to specific sites within the 3'-UTR, miRNAs can decrease gene expression of various mRNAs by either inhibiting translation or directly causing degradation of the transcript. The 3'-UTR also may have silencer regions that bind repressor proteins that inhibit the expression of a mRNA.

The 3'-UTR often contains miRNA response elements (MREs). MREs are sequences to which miRNAs bind. These are prevalent motifs within 3'-UTRs. Among all regulatory motifs within the 3'-UTRs (e.g. including silencer regions), MREs make up about half of the motifs.

As of 2014, the miRBase web site, an archive of miRNA sequences and annotations, listed 28,645 entries in 233 biologic species. Of these, 1,881 miRNAs were in annotated human miRNA loci. miRNAs were predicted to have an average of about four hundred target mRNAs (affecting expression of several hundred genes). Freidman et al. estimate that >45,000 miRNA target sites within human mRNA 3'-UTRs are conserved above background levels, and >60% of human protein-coding genes have been under selective pressure to maintain pairing to miRNAs.

Direct experiments show that a single miRNA can reduce the stability of hundreds of unique mRNAs. Other experiments show that a single miRNA may repress the production of hundreds of proteins, but that this repression often is relatively mild (less than 2-fold).

The effects of miRNA dysregulation of gene expression seem to be important in cancer. For instance, in gastrointestinal cancers, a 2015 paper identified nine miRNAs as epigenetically altered and effective in down-regulating DNA repair enzymes.

The effects of miRNA dysregulation of gene expression also seem to be important in neuropsychiatric disorders, such as schizophrenia, bipolar disorder, major depressive disorder, Parkinson's disease, Alzheimer's disease and autism spectrum disorders.

Regulation of Translation

The translation of mRNA can also be controlled by a number of mechanisms, mostly at the level of initiation. Recruitment of the small ribosomal subunit can indeed be modulated by mRNA secondary structure, antisense RNA binding, or protein binding. In both prokaryotes and eukaryotes, a large number of RNA binding proteins exist, which often are directed to their target sequence by the secondary structure of the transcript, which may change depending on certain conditions, such as temperature or presence of a ligand (aptamer). Some transcripts act as ribozymes and self-regulate their expression.

Examples of Gene Regulation

- Enzyme induction is a process in which a molecule (e.g., a drug) induces (i.e., initiates or enhances) the expression of an enzyme.

- The induction of heat shock proteins in the fruit fly *Drosophila melanogaster*.

- The Lac operon is an interesting example of how gene expression can be regulated.

- Viruses, despite having only a few genes, possess mechanisms to regulate their gene expression, typically into an early and late phase, using collinear systems regulated by anti-terminators (lambda phage) or splicing modulators (HIV).

- GAL4 is a transcriptional activator that controls the expression of GAL1, GAL7, and GAL10 (all of which code for the metabolic of galactose in yeast). The GAL4/UAS system has been used in a variety of organisms across various phyla to study gene expression.

Developmental Biology

A large number of studied regulatory systems come from developmental biology. Examples include:

- The colinearity of the Hox gene cluster with their nested antero-posterior patterning

- It has been speculated that pattern generation of the hand (digits - interdigits) The gradient of Sonic hedgehog (secreted inducing factor) from the zone of polarizing activity in the limb, which creates a gradient of active Gli3, which activates Gremlin, which inhibits BMPs also secreted in the limb, resulting in the formation of an alternating pattern of activity as a result of this reaction-diffusion system.

- Somitogenesis is the creation of segments (somites) from a uniform tissue (Pre-somitic Mesoderm). They are formed sequentially from anterior to posterior. This is achieved in amniotes possibly by means of two opposing gradients, Retinoic acid in the anterior (wavefront) and Wnt and Fgf in the posterior, coupled to an oscillating pattern (segmentation clock) composed of FGF + Notch and Wnt in antiphase.

- Sex determination in the soma of a Drosophila requires the sensing of the ratio of autosomal genes to sex chromosome-encoded genes, which results in the production of sexless splicing factor in females, resulting in the female isoform of doublesex.

Circuitry

Up-regulation and Down-regulation

Up-regulation is a process that occurs within a cell triggered by a signal (originating internal or external to the cell), which results in increased expression of one or more genes and as a result the protein(s) encoded by those genes. Conversely, down-regulation is a process resulting in decreased gene and corresponding protein expression.

- Up-regulation occurs, for example, when a cell is deficient in some kind of receptor. In this case, more receptor protein is synthesized and transported to the membrane of the cell and, thus, the sensitivity of the cell is brought back to normal, reestablishing homeostasis.

- Down-regulation occurs, for example, when a cell is overstimulated by a neurotransmitter, hormone, or drug for a prolonged period of time, and the expression of the receptor protein is decreased in order to protect the cell.

Inducible vs. Repressible Systems

Gene Regulation can be summarized by the response of the respective system:

- Inducible systems - An inducible system is off unless there is the presence of some molecule (called an inducer) that allows for gene expression. The molecule is said to "induce expression". The manner by which this happens is dependent on the control mechanisms as well as differences between prokaryotic and eukaryotic cells.

- Repressible systems - A repressible system is on except in the presence of some molecule (called a corepressor) that suppresses gene expression. The molecule is said to "repress expression". The manner by which this happens is dependent on the control mechanisms as well as differences between prokaryotic and eukaryotic cells.

The GAL4/UAS system is an example of both an inducible and repressible system. GAL4 binds an upstream activation sequence (UAS) to activate the transcription of the GAL1/GAL7/GAL10 cassette. On the other hand, a MIG1 response to the presence of glucose can inhibit GAL4 and therefore stop the expression of the GAL1/GAL7/GAL10 cassette.

Theoretical Circuits

- Repressor/Inducer: an activation of a sensor results in the change of expression of a gene

- negative feedback: the gene product downregulates its own production directly or indirectly, which can result in

  o keeping transcript levels constant/proportional to a factor

  o inhibition of run-away reactions when coupled with a positive feedback loop

  o creating an oscillator by taking advantage in the time delay of transcription and translation, given that the mRNA and protein half-life is shorter

- positive feedback: the gene product upregulates its own production directly or indirectly, which can result in

  o signal amplification

  o bistable switches when two genes inhibit each other and both have positive feedback

  o pattern generation

Study Methods

In general, most experiments investigating differential expression used whole cell extracts of RNA, called steady-state levels, to determine which genes changed and by how much they did. These are, however, not informative of where the regulation has occurred and may actually mask conflicting regulatory processes, but it is still the most commonly analysed (quantitative PCR and DNA microarray).

When studying gene expression, there are several methods to look at the various stages. In eukaryotes these include:

- The local chromatin environment of the region can be determined by ChIP-chip analysis by pulling down RNA Polymerase II, Histone 3 modifications, Tritho-rax-group protein, Polycomb-group protein, or any other DNA-binding element to which a good antibody is available.

- Epistatic interactions can be investigated by synthetic genetic array analysis

- Due to post-transcriptional regulation, transcription rates and total RNA levels differ significantly. To measure the transcription rates nuclear run-on assays can be done and newer high-throughput methods are being developed, using thiol labelling instead of radioactivity.

- Only 5% of the RNA polymerised in the nucleus actually exits, and not only introns, abortive products, and non-sense transcripts are degradated. Therefore, the differences in nuclear and cytoplasmic levels can be see by separating the two fractions by gentle lysis.

- Alternative splicing can be analysed with a splicing array or with a tiling array.

- All in vivo RNA is complexed as RNPs. The quantity of transcripts bound to specific protein can be also analysed by RIP-Chip. For example, DCP2 will give an indication of sequestered protein; ribosome-bound gives and indication of transcripts active in transcription (although it should be noted that a more dated method, called polysome fractionation, is still popular in some labs)

- Protein levels can be analysed by Mass spectrometry, which can be compared only to quantitative PCR data, as microarray data is relative and not absolute.

- RNA and protein degradation rates are measured by means of transcription inhibitors (actinomycin D or α-amanitin) or translation inhibitors (Cyclohexi-mide), respectively.

Need for Gene Regulation

The biochemical processes and signalling inside cells is mostly enzyme driven and are energy dependent. The enzymatic proteins which affect these processes are products of gene transcription and translation. These genes are switched on or off depend on the requirements of the time scales.

Gene switching happens due to a variety of reasons, some of which are listed below:

1. Selective switching of genes prevents wastage of energy and cellular resources. Many genes posses the natural capability to switch ON or switch OFF genes based on their requirements

2. Gene switching events are one of the key processes required at development phases of multi cellular organisms during cell fate decisions

3. Unregulated gene expression leads to diseases for example, the onset of cancer can be attributed to aberrant gene expression pattern where cancer cells lose their ability to regulate mitosis-an uncontrolled cell division.

How do External Signals Control Gene Expression?

Extra cellular cues from the environment like temperature, changes in pH and changes in internal biochemical signalling have been shown to alter gene expression to a significant extent. The response of a liver cell to the steroid- gluoco corticoid hormone is a striking example of altered gene expression. Gluoco corticoid hormones are released in the body during starvation or during extreme activity of exercising. Gluoco corticoid signalling in liver enhances glucose production and alters expression of tyrosine amino transferase which aids the conversion from tyrosine to glucose. In the absence of the hormone, basal levels of these enzymes are maintained at basic levels. Adipose cells respond differently to gluoco corticoids. When a fat cell is exposed to gluoco corticoid, the levels of tyrosine amino transferases are reduced. Some of the cell types do not respond to gluoco corticoid at all. We therefore understand that the same extracellular signal elicits different responses from different cell types.

Gene Switching

A differentiated cell example neuronal cell performs its unique and specific function by switching ON a specific set of genes and by switching OFF another. Some of these genes are permanently active (switched ON) because they carry out vital functions of the cell and because they contain the blueprint for vital enzymes required for cellular function. Certain genes not required for a particular function in the cell may be switched OFF. The production of insulin in pancreatic cell signals the ON state of the insulin gene while other genes irrelevant for insulin signalling may be switched OFF.

The level of expression of a specific gene in a cell is a major factor in the expression control mechanism. For example, the amount of melanin expressed by the skin cell decides the color of the skin. Melanin production is under the type control of two genes each of which show dominant and recessive expression as represented in Figure.

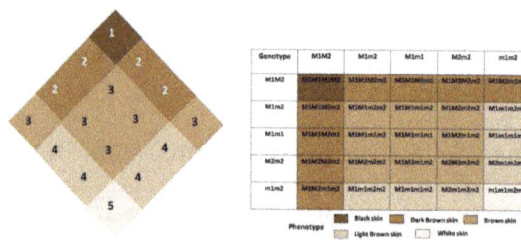

Genotypes and phenotypes expressed by 2 melanin genes (dominant and recessive)

The Figure illustrates 5 possible genotypes that the melanin gene can express. We find that 4 dominant alleles represent enhanced expression of melanin leading to the phenotype of black skin.

Gene Control in Eukaryotic Systems

Studying gene control in eukaryotic cells is an active and attractive pursuit for experimental biologists. The term 'Eukaryotic Gene Control' is not restricted to the process of transcribing to mRNA. For example, the transcription units for mRNAs do not mark the final mRNA product in the case of vertebrate cells. Several control points exists from the processing of primary RNA transcripts in the cell nucleus till the mRNA is translated in the cytoplasm. Specifically the eukaryotic genes undergo more levels of regulation than the bacterial genes. Transcription regulation, RNA processing in the nucleus, mRNA stability, and the frequency of translation events in the cytoplasm are the vital steps at which eukaryotic genes exercise control in expression. The same has been illustrated in Figure.

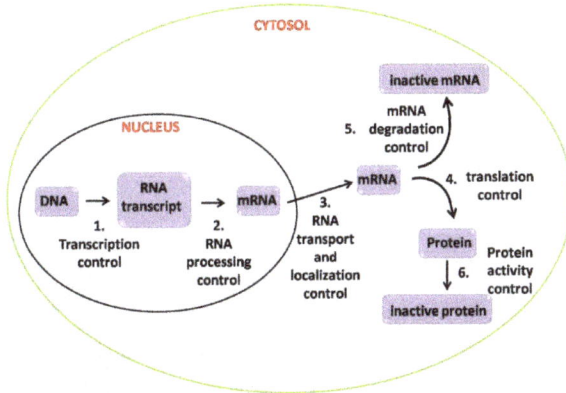

Gene regulatory pathways extending from the nucleus
to the cytoplasm of a typical eukaryotic cell

The transcription process in eukaryotic cells is more complex than in the prokaryotic systems. Gene regulation in these systems can be effected in two ways as shown in the Figure.

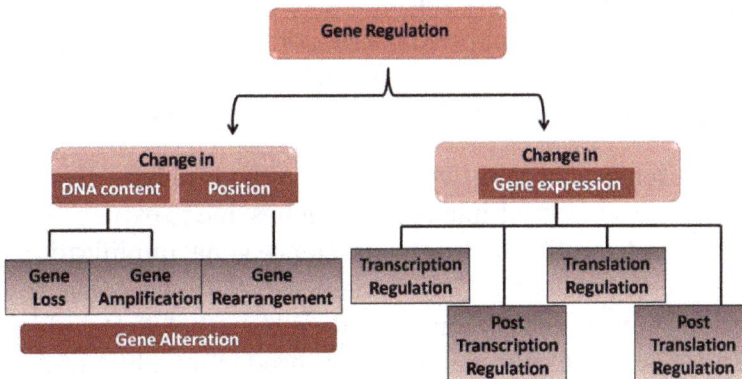

Mechanisms of gene regulation in eukaryotes

The regulation could be effected through

1   Control mechanisms that involve change in DNA content or position. Here the DNA of somatic cells can be altered but these changes do not affect gametes so that they are not inherited by the offspring.

2   Control mechanisms involving changes in expression patterns. Regulatory mechanisms involving change in DNA content or position involves alterations at the level of the gene and may be due to

    (a)   Gene loss

    (b)   Gene amplification

    (c)   Gene rearrangement

Regulatory mechanisms involving changes in expression pattern include

    (a)   Transcriptional control

    (b)   Post-transcriptional control

    (c)   Translational control

    (d)   Post-translational control

Gene alteration is one of the key mechanisms of gene regulation in eukaryotes. This may be carried out through mechanisms like gene loss, gene amplification or through gene rearrangement. Gene alteration plays a role in regulating gene expression under developmental stages, drug resistance etc. Gene loss involves elimination of the gene from differentiating cells in protozoans as well as in metazoans. Lineage specific gene loss is a major evolutionary process studied from gene sets of completely sequenced genomes. In prokaryotes, gene loss and horizontal gene transfer contributes to an intensive gene flux that shapes the genomes of these organisms. In eukaryotes lineage specific gene loss is greater than in prokaryotes. A very strong selection pressures result in dramatic gene loss. For example, Encephalitozoon cuniculi is a eukaryotic intracellular parasite, which now has been estimated to have approximately 2000 genes compared to around 6000 genes in the genomes of yeast, representing a striking case of massive gene loss. Complete chromosomes are preserved only in cells serving as precursors for gamete formation.

Gene amplification refers to a selective increase in the number of copies of a gene sequence. Cancer cells produce multiple copies of genes due to extra cellular signalling or in response to signals from the environment. Hence gene amplification occurs during malignancy and in certain cases cells become resistant to drugs due to the process of gene amplification. Cancer cells when treated with methotrexate which blocks dihydrofolate reductase undergo gene amplification producing dihydrofolate reductase in abundant amount to circumvent the effect of methotrexate. The term 'gene amplification' is often used to refer to the laboratory technique called 'polymerase Chain Reaction' which enables amplifying gene sequences in an eppendorf tube.

Gene rearrangement involves the translocation of a gene from one region of a genome to another. This involves rearrangement of substantial stretches of DNA in the somatic cells and can alter gene expression. Transposons are DNA stretches that move across locations within the genome. These transposons control gene expression depending on their location. For example, if the transposon is inserted in a transcription regulatory sequence it regulates the levels of protein production. It may also carry a gene that is activated if it is inserted downstream of a promoter sequence. Mammalian systems such as mouse or human being synthesize large amounts of specific antibodies against any foreign antigens on exposure. Antibodies are composed of two heavy and two light chains. The specificity of the antibody is determined by the amino acid sequences of the variable regions of both the light and the heavy chains. The rearrangement of various immunoglobulin genes leads to the formation of a variety of antibodies specific for diverse antigens. African Trypanosomes sequentially express various surface antigens in their mammalian host. The expression of such Variable Surface Glycoprotein (VSG) antigens has been shown to be related to DNA rearrangements. (Williams R. et al., Nature 1979, Payes et al. PNAS 1981). Trypanosomes circumvent host defense by replacing a copy of VSG (Variant Surface Glycoprotein) gene at the expression site by another VSG gene at some other site.

We shall discuss in the next class how changes in gene expression can be brought about at the level of transcription, translation and at the level of post transcription and post translation through precise regulatory mechanisms that control expressions.

Gene Control Mechanisms at Transcriptional and Post transcriptional Levels

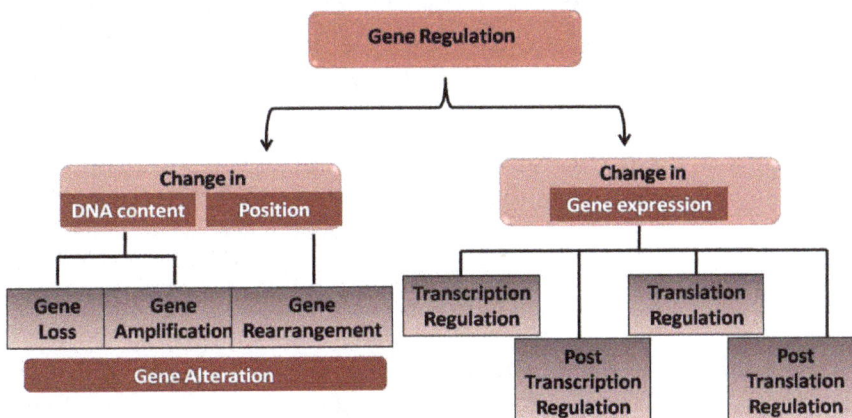

Mechanisms of gene regulation in eukaryotes

The complicated mechanisms of mRNA formation, the complex structures of the nucleus and cytoplasm of eukaryotic cells, the robust genetic program involved in differentiation of eukaryotic cells strikingly differentiate eukaryotic gene control from bacterial gene regulation. As stated earlier, the eukaryotic genes undergo several more steps of regulation than the bacterial genes. Single molecule experiments have clearly and accurately decipher the mechanical steps involved in eukaryotic gene control. Such experiments have even determined at the transcription level, the amount of specific mRNA molecules that are regulated, the processing of the nuclear RNA and the stability of mRNA. As stated in the previous class, gene regulation can be brought about by altering gene expression states in a cell. This can be effected at the levels of transcription and translation and during post transcription and post translational events.

Transcriptional Regulation

In molecular biology and genetics, transcriptional regulation is the means by which a cell regulates the conversion of DNA to RNA (transcription), thereby orchestrating gene activity. A single gene can be regulated in a range of ways, from altering the number of copies of RNA that are transcribed, to the temporal control of when the gene is transcribed. This control allows the cell or organism to respond to a variety of intra- and extracellular signals and thus mount a response. Some examples of this include producing the mRNA that encode enzymes to adapt to a change in a food source, producing the gene products involved in cell cycle specific activities, and producing the gene products responsible for cellular differentiation in higher eukaryotes, as studied in evolutionary developmental biology.

The regulation of transcription is a vital process in all living organisms. It is orchestrated by transcription factors and other proteins working in concert to finely tune the amount of RNA being produced through a variety of mechanisms. Prokaryotic organisms and eukaryotic organisms have very different strategies of accomplishing control over transcription, but some important features remain conserved between the two. Most importantly is the idea of combinatorial control, which is that any given gene is likely controlled by a specific combination of factors to control transcription. In a hypothetical example, the factors A and B might regulate a distinct set of genes from the combination of factors A and C. This combinatorial nature extends to complexes of far more than two proteins, and allows a very small subset (less than 10%) of the genome to control the transcriptional program of the entire cell.

Transcription Regulation in Prokaryotes

Much of the early understanding of transcription came from prokaryotic organisms, although the extent and complexity of transcriptional regulation is greater in eukaryotes. Prokaryotic transcription is governed by three main sequence elements:

- Promoters are elements of DNA that may bind RNA polymerase and other proteins for the successful initiation of transcription directly upstream of the gene.

- Operators recognize repressor proteins that bind to a stretch of DNA and inhibit the transcription of the gene.

- Positive control elements that bind to DNA and incite higher levels of transcription

While these means of transcriptional regulation also exist in eukaryotes, the transcriptional landscape is significantly more complicated both by the number of proteins involved as well as by the presence of introns and the packaging of DNA into histones.

The transcription of a basic prokaryotic gene is dependent on the strength of its promoter and the presence of activators or repressors. In the absence of other regulatory elements, a promoter's sequence-based affinity for RNA polymerases varies, which results in the production of different amounts of transcript. The variable affinity of RNA polymerase for different promoter sequences is related to regions of consensus sequence upstream of the transcription start site. The more nucleotides of a promoter that agree with the consensus sequence, the stronger the affinity of the promoter for RNA Polymerase likely is.

In the absence of other regulatory elements, the default state of a prokaryotic transcript is to be in the "on" configuration, resulting in the production of some amount of transcript. This means that transcriptional regulation in the form of protein repressors and positive control elements can either increase or decrease transcription. Repressors often physically occupy the promoter location, occluding RNA polymerase from binding. Alternatively a repressor and polymerase may bind to the DNA at the same time with a physical interaction between the repressor preventing the opening of the DNA for access to the minus strand for transcription. This strategy of control is distinct from eukaryotic transcription, whose basal state is to be off and where co-factors required for transcription initiation are highly gene dependent.

Sigma factors are specialized bacterial proteins that bind to RNA polymerases and orchestrate transcription initiation. Sigma factors act as mediators of sequence-specific transcription, such that a single sigma factor can be used for transcription of all housekeeping genes or a suite of genes the cell wishes to express in response to some external stimuli such as stress.

Eukaryotic Transcription Regulation

The added complexity of generating a eukaryotic cell carries with it an increase in the complexity of transcriptional regulation. Eukaryotes have three RNA polymerases, known as Pol I, Pol II, and Pol III. Each polymerase has specific targets and activities, and is regulated by independent mechanisms. There are a number of additional mechanisms through which polymerase activity can be controlled. These mechanisms can be generally grouped into three main areas:

- Control over polymerase access to the gene. This is perhaps the broadest of the three control mechanisms. This includes the functions of histone remodeling enzymes, transcription factors, enhancers and repressors, and many other complexes

- Productive elongation of the RNA transcript. Once polymerase is bound to a promoter, it requires another set of factors to allow it to escape the promoter complex and begin successfully transcribing RNA.

- Termination of the polymerase. A number of factors which have been found to control how and when termination occurs, which will dictate the fate of the RNA transcript.

All three of these systems work in concert to integrate signals from the cell and change the transcriptional program accordingly.

While in prokaryotic systems the basal transcription state can be thought of as nonrestrictive (that is, "on" in the absence of modifying factors), eukaryotes have a restrictive basal state which requires the recruitment of other factors in order to generate RNA transcripts. This difference is largely due to the compaction of the eukaryotic genome by winding DNA around histones to form higher order structures. This compaction makes the gene promoter inaccessible without the assistance of other factors in the nucleus, and thus chromatin structure is a common site of regulation. Similar to the sigma factors in prokaryotes, the general transcription factors (GTFs) are a set of factors in eukaryotes that are required for all transcription events. These factors are responsible for stabilizing binding interactions and opening the DNA helix to allow the RNA polymerase to access the template, but generally lack specificity for different promoter sites. A large part of gene regulation occurs through transcription factors that either recruit or inhibit the binding of the general transcription machinery and/or the polymerase. This can be accomplished through close interactions with core promoter elements, or through the long distance enhancer elements.

Once a polymerase is successfully bound to a DNA template, it often requires the assistance of other proteins in order to leave the stable promoter complex and begin elongating the nascent RNA strand. This process is called promoter escape, and is another step at which regulatory elements can act to accelerate or slow the transcription process. Similarly, protein and nucleic acid factors can associate with the elongation complex and modulate the rate at which the polymerase moves along the DNA template.

Regulation at the Level of Chromatin State

In eukaryotes, genomic DNA is highly compacted in order to be able to fit it into the nucleus. This is accomplished by winding the DNA around protein octamers called histones, which has consequences for the physical accessibility of parts of the genome at any given time. Significant portions are silenced through histone modifications, and thus are inaccessible to the polymerases or their cofactors. The highest level of transcription regulation occurs through the rearrangement of histones in order to expose or sequester genes, because these processes have the ability to render entire regions of a chromosome inaccessible such as what occurs in imprinting.

Histone rearrangement is facilitated by post-translational modifications to the tails of the core histones. A wide variety of modifications can be made by enzymes such as the histone acetyltransferases (HATs), histone methyltransferases (HMTs), and histone deacetylases (HDACs), among others. These enzymes can add or remove covalent modifications such as methyl groups, acetyl groups, phosphates, and ubiquitin. Histone modifications serve to recruit other proteins which can either increase the compaction of the chromatin and sequester promoter elements, or to increase the spacing between histones and allow the association of transcription factors or polymerase on open DNA. For example, H3K27 trimethylation by the polycomb complex PRC2 causes chromosomal compaction and gene silencing. These histone modifications may be created by the cell, or inherited in an epigenetic fashion from a parent.

Regulation through Transcription Factors and Enhancers

Transcription Factors

Transcription factors are proteins that bind to specific DNA sequences in order to regulate the expression of a given gene. The power of transcription factors resides in their ability to activate and/or repress wide repertoires of downstream target genes. The fact that these transcription factors work in a combinatorial fashion means that only a small subset of an organism's genome encodes transcription factors. Transcription factors function through a wide variety of mechanisms. Often they are at the end of a signal transduction pathway that functions to change something about the factor, like its subcellular localization or its activity. Post-translational modifications to transcription factors located in the cytosol can cause them to translocate to the nucleus where they can interact with their corresponding enhancers. Others are already in the nucleus, and are modified to enable the interaction with partner transcription factors. Some post-translational modifications known to regulate the functional state of transcription factors are phosphorylation, acetylation, SUMOylation and ubiquitylation. Transcription factors can be divided in two main categories: activators and repressors. While activators can interact directly or indirectly with the core machinery of transcription through enhancer binding, repressors predominantly recruit co-repressor complexes leading to transcriptional repression by chromatin condensation of enhancer regions. It may also happen that a repressor may function by allosteric competition against a determined activator to repress gene expression: overlapping DNA-binding motifs for both activators and repressors induce a physical competition to occupy the site of binding. If the repressor has a higher affinity for its motif than the activator, transcription would be effectively blocked in the presence of the repressor. Tight regulatory control is achieved by the highly dynamic nature of transcription factors. Again, many different mechanisms exist to control whether a transcription factor is active. These mechanisms include control over protein localization or control over whether the protein can bind DNA. An example of this is the protein HSF1, which remains bound to Hsp70 in the cytosol and is only translocated into the nucleus upon cellular stress such as heat shock. Thus the genes under the control of this transcription factor will remain untranscribed unless the cell is subjected to stress.

Enhancers

Enhancers or cis-regulatory modules/elements (CRM/CRE) are non-coding DNA sequences containing multiple activator and repressor binding sites. Enhancers range from 200 bp to 1 kb in length and can be either proximal, 5' upstream to the promoter or within the first intron of the regulated gene, or distal, in introns of neighboring genes or intergenic regions far away from the locus. Through DNA looping, active enhancers contact the promoter dependently of the core DNA binding motif promoter specificity. Promoter-enhancer dichotomy provides the basis for the functional interaction between transcription factors and transcriptional core machinery to trigger RNA Pol II escape from the promoter. Whereas one could think that there is a 1:1 enhancer-promoter ratio, studies of the human genome predict that an active promoter interacts with 4 to 5 enhancers. Similarly, enhancers can regulate more than one gene without linkage restriction and are said to "skip" neighboring genes to regulate more distant ones. Even though infrequent, transcriptional regulation can involve elements located in a chromosome different to one where the promoter resides. More interestingly, proximal enhancers or promoters of neighboring genes can serve as platforms to recruit more distal elements.

Regulatory Landscape

Transcriptional initiation, termination and regulation are mediated by "DNA looping" which brings together promoters, enhancers, transcription factors and RNA processing factors to accurately regulate gene expression. Chromosome conformation capture (3C) and more recently Hi-C techniques provided evidence that active chromatin regions are "compacted" in nuclear domains or bodies where transcriptional regulation is enhanced24614317. The configuration of the genome is essential for enhancer-promoter proximity. Cell-fate decisions are mediated upon highly dynamic genomic reorganizations at interphase to modularly switch on or off entire gene regulatory networks through short to long range chromatin rearrangements. Related studies demonstrate that metazoan genomes are partitioned in structural and functional units around a megabase long called Topological association domains (TADs) containing dozens of genes regulated by hundreds of enhancers distributed within large genomic regions containing only non-coding sequences. The function of TADs is to regroup enhancers and promoters interacting together within a single large functional domain instead of having them spread in different TADs. However, studies of mouse development point out that two adjacent TADs may regulate the same gene cluster. The most relevant study on limb evolution shows that the TAD at the 5' of the HoxD gene cluster in tetrapod genomes drives its expression in the distal limb bud embryos, giving rise to the hand, while the one located at 3' side does it in the proximal limb bud, giving rise to the arm. Still, it is not known whether TADs are an adaptive strategy to enhance regulatory interactions or an effect of the constrains on these same interactions. TAD boundaries are often composed by housekeeping genes, tRNAs, other highly expressed sequences and Short Interspersed Elements (SINE). While these genes may take advantage of their border position to be ubiquitously expressed, they are not directly linked with TAD edge formation. The specific molecules identified at boundaries of TADs are called insulators or architectural proteins because they not only block enhancer leaky expression but also ensure an accurate compartmentalization of cis-regulatory inputs to the targeted promoter. These insulators are DNA-binding proteins like CTCF and TFIIIC that help recruiting structural partners such as cohesins and condensins. The localization and binding of architectural proteins to their corresponding binding sites is regulated by post-translational modifications. Interestingly, DNA binding motifs recognized by architectural proteins are either of high occupancy and at around a megabase of each other or of low occupancy and inside TADs. High occupancy sites are usually conserved and static while intra-TADs sites are dynamic according to the state of the cell therefore TADs themselves are compartmentalized in subdomains that can be called subTADs from few kb up to a TAD long (19). When architectural binding sites are at less than 100 kb from each other, Mediator proteins are the architectural proteins cooperate with cohesin. For subTADs larger than 100 kb and TAD boundaries, CTCF is the typical insulator found to interact with cohesion.

Regulation of the Pre-initiation Complex and Promoter Escape

In eukaryotes, ribosomal rRNA and the tRNAs involved in translation are controlled by RNA polymerase I (Pol I) and RNA polymerase III (Pol III). RNA Polymerase II is responsible for the production of messenger RNA (mRNA) within the cell. Particularly for Pol II, much of the regulatory checkpoints in the transcription process occur in the assembly and escape of the pre-initiation complex. A gene-specific combination of transcription factors will recruit TFIID and/or TFIIA to the core promoter, followed by the association of TFIIB, creating a stable complex onto which the rest of the General Transcription Factors (GTFs) can assemble. This complex is relatively stable, and can undergo multiple rounds of transcription initiation. After the binding of TFIIB and TFIID, Pol II the rest of the GTFs can assemble. This assembly is marked by the post-translational modification (typically phosphorylation) of the C-terminal domain (CTD) of Pol II through a number of kinases. The CTD is a large, unstructured domain extending from the RbpI subunit of Pol II, and consists of many repeats of the heptad sequence YSPTSPS. TFIIH, the helicase that remains associated with Pol II throughout transcription, also contains a subunit with kinase activity which will phosphorylate the serines 5 in the heptad sequence. Similarly, both CDK8 (a subunit of the massive multiprotein Mediator complex) and CDK9 (a subunit of the p-TEFb elongation factor), have kinase activity towards other residues on the CTD. These phosphorylation events promote the transcription process and serve as sites of recruitment for mRNA processing machinery. All three of these kinases respond to upstream signals, and failure to phosphorylate the CTD can lead to a stalled polymerase at the promoter.

Regulation of Transcription in Cancer

In vertebrates, the majority of gene promoters contain a CpG island with numerous CpG sites. When many of a gene's promoter CpG sites are methylated the gene becomes silenced. Colorectal cancers typically have 3 to 6 driver mutations and 33 to 66 hitchhiker or passenger mutations. However, transcriptional silencing may be of more importance than mutation in causing progression to cancer. For example, in colorectal cancers about 600 to 800 genes are transcriptionally silenced by CpG island methylation. Transcriptional repression in cancer can also occur by other epigenetic mechanisms, such as altered expression of microRNAs. In breast cancer, transcriptional repression of BRCA1 may occur more frequently by over-expressed microRNA-182 than by hypermethylation of the BRCA1 promoter.

Transcription in eukaryotic cells requires at the first level of control the choice of RNA polymerases. Eukaryotes have three different RNA polymerases, polymerases I, II and III. In vivo and in vitro experiments have shown that the RNA polymerase II initiates transcription from a site located near a conserved 8 to 10 nucleotide region, 25 to 30 nucleotides upstream (5') from the RNA start site. The nucleotides TATA which are strongly conserved at this site gives the name TATA box or Goldberg-Hogness box. There are other sequences equally important in regulating the access polymerases to the DNA.

Transcription factors play a key role in helping RNA polymerase to bind to promoter region. Transcription factors bind to various consensus sequences like TATA box in Figure. TATA binding protein recognises this box and recruits RNA polymerase II to promoter site for transcription initiation. Transcription factors called activators bind to enhancers. Enhancers lie far from the transcription start site and are position insensitive, while the promoter sites lie close to transcription start site and are position sensitive. Genes lacking TATA box have initiator sequence for initiating transcription. Combination of specific transcription factors is critical for the transcription of a particular gene. Insulators are DNA sequences regulating gene expression. They play a key role in enhancer blocking. Positional enhancer blocking occurs only when insulator lies in between enhancer and operator. Insulators allow genes to remain independent in the genome by restraining the unnecessary signals influencing the gene.

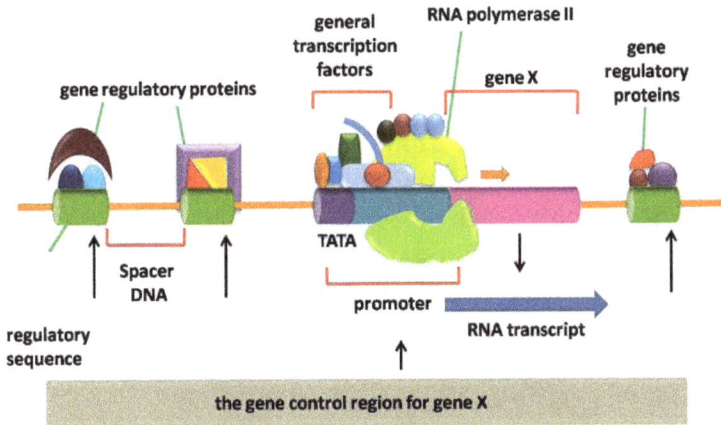

Factors that influence the regulation of a typical gene X

A variety of structural motifs present in transcription factors interact with DNA at specific sequences. On the basis of DNA binding domains they accommodate, eukaryotic transcription factors are classified as

Zinc Finger proteins which have regions folding around a central Zn^{2+} ion. Generally DNA binding domains of transcription factors contain such regions. Proteins which do not bind to DNA also show Zn finger patterns.

Leucine Zipper proteins have a leucine at every seventh position on the protein sequence. Such proteins function as transcription factors when present in the form of a dimer. The presence of Leucine helps in parallel zippering of dimer. Yeast Gcn4 is an example of such protein.

Helix turn helix motif is composed of two alpha helices connected by a short amino acid strand. One of them is an amino terminal helix and the other a C- terminal helix. The C-terminal helix is critical for DNA recognition while the other helix stabilizes the DNA- protein interaction. The C-terminal helix binds DNA at major grooves through hydrogen bonding. The Cro, lambda repressor and Catabolite gene activator protein are examples of Helix turn helix motif structures.

Hormones play a crucial role in gene transcription. Type 1 hormones like gluco corticoids being hydrophobic bind to intracellular receptors located in cytoplasm. Hormone receptor interaction causes dissociation of the nuclear localisation signal of receptor which is then transported to nucleus where it binds to hormone response elements to cause transcription of gene. Type II hormones like epinephrine being hydrophilic bind to cell surface receptors and activate G- proteins, leading to gene transcription.

Post-transcriptional Regulation

Post-transcriptional regulation is the control of gene expression at the RNA level, therefore between the transcription and the translation of the gene.

Mechanism

After being produced, the stability and distribution of the different transcripts is regulated (post-transcriptional regulation) by means of RNA binding protein (RBP) that control the various steps and rates cripts: events such as alternative splicing, nuclear degradation (exosome), processing, nuclear export (three alternative pathways), sequestration in P-bodies for storage or degradation and ultimately translation. These proteins achieve these events thanks to a RNA recognition motif (RRM) that binds a specific sequence or secondary structure of the transcripts, typically at the 5' and 3' UTR of the transcript.

Modulating the capping, splicing, addition of a Poly(A) tail, the sequence-specific nuclear export rates and in several contexts sequestration of the RNA transcript occurs in eukaryotes but not in prokaryotes. This modulation is a result of a protein or transcript which in turn is regulated and may have an affinity for certain sequences.

- Capping changes the five prime end of the mRNA to a three prime end by 5'-5' linkage, which protects the mRNA from 5' exonuclease, which degrades foreign RNA. The cap also helps in ribosomal binding.

- Splicing removes the introns, noncoding regions that are transcribed into RNA, in order to make the mRNA able to create proteins. Cells do this by spliceosomes binding on either side of an intron, looping the intron into a circle and then cleaving it off. The two ends of the exons are then joined together.

- Addition of poly(A) tail otherwise known as polyadenylation. That is, a stretch of RNA that is made solely of adenine bases is added to the 3' end, and acts as a buffer to the 3' exonuclease in order to increase the half life of mRNA. In addition, a long poly(A) tail can increase translation. Poly(A)-binding protein (PABP) binds to a long poly(A) tail and mediates the interaction between EIF4E and EIF4G which encourages the initiation of translation.

- RNA editing is a process which results in sequence variation in the RNA molecule, and is catalyzed by enzymes. These enzymes include the Adenosine Deaminase Acting on RNA (ADAR) enzymes, which convert specific adenosine residues to inosine in an mRNA molecule by hydrolytic deamination. Three ADAR enzymes have been cloned, ADAR1, ADAR2 and ADAR3, although only the first two subtypes have been shown to have RNA editing activity. Many mRNAs are vulnerable to the effects of RNA editing, including the glutamate receptor subunits GluR2, GluR3, GluR4, GluR5 and GluR6 (which are components of the AMPA and kainate receptors), the serotonin2C receptor, the GABA-alpha3 receptor subunit, the tryptophan hydroxlase enzyme TPH2, the hepatitis delta virus and more than 16% of microRNAs. In addition to ADAR enzymes, CDAR enzymes exist and these convert cytosines in specific RNA molecules, to uracil. These enzymes are termed 'APOBEC' and have genetic loci at 22q13, a region close to the chromosomal deletion which occurs in velocardiofacial syndrome (22q11) and which is linked to psychosis. RNA editing is extensively studied in relation to infectious diseases, because the editing process alters viral function.

- mRNA Stability can be manipulated in order to control its half-life, and the poly(A) tail has some effect on this stability, as previously stated. Stable mRNA can have a half life of up to a day or more which allows for the production of more protein product; unstable mRNA is used in regulation that must occur quickly.

microRNA Mediated Regulation

MicroRNAs (miRNAs) appear to regulate the expression of more than 60% of protein coding genes of the human genome. If an miRNA is abundant it can behave as a "switch", turning some genes on or off. However, altered expression of many miRNAs only leads to a modest 1.5- to 4-fold change in protein expression of their target genes. Individual miRNAs often repress several hundred target genes. Repression usually occurs either through translational silencing of the mRNA or through degradation of the mRNA, via complementary binding, mostly to specific sequences in the 3' untranslated region of the target gene's mRNA. The mechanism of translational silencing or degradation of mRNA is implemented through the RNA-induced silencing complex (RISC).

Feedback in the Regulation of RNA Binding Proteins

In metazoans and bacteria, many genes involved in post-post transcriptional regulation are regulated post transcriptionally. For Drosophila RBPs associated with splicing or nonsense mediated decay, analyses of protein-protein and protein-RNA interaction profiles have revealed ubiquitous interactions with RNA and protein products of the same gene. It remains unclear whether these observations are driven by ribosome proximal or ribosome mediated contacts, or if some protein complexes, particularly RNPs, undergo co-translational assembly.

Significance

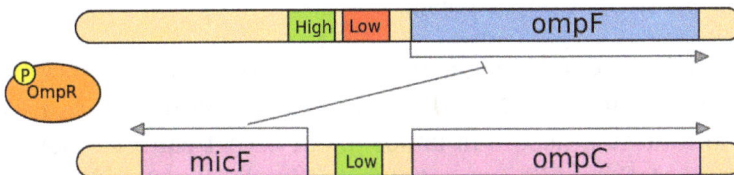

A prokaryotic example: *Salmonella enterica* (a pathogenic γ-proteobacterium) can express two alternative porins depending on the external environment (gut or murky water), this system involves EnvZ (osomotic sensor) which activates OmpR (transcription factor) which can bind to a high affinity promoter even at low concentrations and the low affinity promoter only at high concentrations (by definition): when the concentration of this transcription factor is high it activates OmpC and micF and inhibits OmpF, OmpF is further inhibited post-transcriptionally by micF RNA which binds to the OmpF transcript

This area of study has recently gained more importance due to the increasing evidence that post-transcriptional regulation plays a larger role than previously expected. Even though protein with DNA binding domains are more abundant than protein with RNA binding domains, a recent study by Cheadle et al. (2005) showed that during T-cell activation 55% of significant changes at the steady-state level had no corresponding changes at the transcriptional level, meaning they were a result of stability regulation alone.

Furthermore, RNA found in the nucleus is more complex than that found in the cytoplasm: more than 95% (bases) of the RNA synthesized by RNA polymerase II never reaches the cytoplasm. The main reason for this is due to the removal of introns which account for 80% of the total bases. Some studies have shown that even after processing the levels of mRNA between the cytoplasm and the nucleus differ greatly.

Developmental biology is a good source of models of regulation, but due to the technical difficulties it was easier to determine the transcription factor cascades than regulation at the RNA level. In fact several key genes such as nanos are known to bind RNA but often their targets are unknown. Although RNA binding proteins may regulate post transcriptionally large amount of the transcriptome, the targeting of a single gene is of interest to the scientific community for medical reasons, this is RNA interference and microRNAs which are both examples of posttranscriptional regulation, which regulate the destruction of RNA and change the chromatin structure. To study post-transcriptional regulation several techniques are used, such as RIP-Chip (RNA immunoprecipitation on chip).

microRNA Role in Cancer

Deficiency of expression of a DNA repair gene occurs in many cancers. Altered microRNA (miRNA) expression that either decreases accurate DNA repair or increases inaccurate microhomology-mediated end joining (MMEJ) DNA repair is often observed in cancers. Deficiency of accurate DNA repair may be a major source of the high frequency of mutations in cancer. Repression of DNA repair genes in cancers by changes in the levels of microRNAs may be a more frequent cause of repression than mutation or epigenetic methylation of DNA repair genes.

For instance, BRCA1 is employed in the accurate homologous recombinational repair (HR) pathway. Deficiency of BRCA1 can cause breast cancer. Down-regulation of BRCA1 due to mutation occurs in about 3% of breast cancers. Down-regulation of BRCA1 due to methylation of its promoter occurs in about 14% of breast cancers. However, increased expression of miR-182 down-regulates BRCA1 mRNA and protein expression, and increased miR-182 is found in 80% of breast cancers.

In another example, a mutated constitutively (persistently) expressed version of the oncogene c-Myc is found in many cancers. Among many functions, c-Myc negatively regulates microRNAs miR-150 and miR-22. These microRNAs normally repress expression of two genes essential for MMEJ, Lig3 and Parp1, thereby inhibiting this inaccurate, mutagenic DNA repair pathway. Muvarak et al. showed, in leukemias, that constitutive expression of c-Myc, leading to down-regulation of miR-150 and miR-22, allowed increased expression of Lig3 and Parp1. This generates genomic instability through increased inaccurate MMEJ DNA repair, and likely contributes to progression to leukemia.

To show the frequent ability of microRNAs to alter DNA repair expression, Hatano et al. performed a large screening study, in which 810 microRNAs were transfected into cells that were then subjected to ionizing radiation (IR). For 324 of these microRNAs, DNA repair was reduced (cells were killed more efficiently by IR) after transfection. For a further 75 microRNAs, DNA repair was increased, with less cell death after IR. This indicates that alterations in microRNAs may often down-regulate DNA repair, a likely important early step in progression to cancer.

Post Transcriptional Control

Post transcriptional control comes into play after the primary transcript is produced. Alternative mRNA splicing is one of the post transcriptional regulatory methods and follows tissue specific or developmental fashion. Tissue specific regulation results in the formation of two different proteins from the same gene in different cells while differentiation specific regulation results in the formation of more than one functionally different protein in the same cell. The best example is the production of calcitonin in thyroid cells, but calcitonin related peptide in brain cells due to alternative splicing. Alternative splicing results in different combinations of exons producing different proteins from the same transcript. Alternative splicing has a switching function which leads to functional or inactive proteins.

Translational Control

This mechanism provides hindrance to protein synthesis through proteins involved in translation. This includes stability of mRNA, probability of translation initiation apart from regulation of overall protein synthesis. These mechanisms control the amount of protein produced from mRNA during a translational event.

Stability of mRNA is important step that controls protein synthesis. Usually the eukaryotic mRNA survives for 3 hours but in certain cases where the amount of protein required in higher amounts, the mRNA survives for several days. In silk worms which produce fibroin, the amount of fibroin for cocoon formation takes days thus the fibroin mRNA survives for several days. Repeated translation of mRNA results in the formation of 10^5 fibroin molecules from 10^4 fibroin mRNAs. Oviduct cells synthesize ovalbumin in chickens but such cells have single copy of this gene per haploid dose. Thus in such circumstances the lifetime of mRNA should be high so that same mRNA is translated repeatedly to achieve the desired protein level.

Reticulocytes synthesize a single type of protein. In such circumstance translation is regulated by modulating total protein synthesis. Globin synthesis in rabbit reticulosites is controlled at the level of translation. Such cells lack nucleus and thus cannot be controlled at the transcriptional level. The process of translation requires elongation factor eIF2 and eIF2 stimulating protein (ESP) besides other molecular players that regulate the event.

Post Translational Control

This control operates after the protein is synthesised in the cell. Post translational modifications include covalent modifications which occur in proteins after translation. Such modifications add another layer of complexity to the complexity embedded in the genome. Post translational modifications play a critical role in regulating activity, localization, and enabling inter molecular interactions inside the cell.

Glysosylation regulates protein stability, influences protein folding kinetics. Ubiquitylation of proteins acts as a signal for proteasomal degradation of proteins. Ubiquitin is a composed of 76 amino acid residues and links to proteins via isopeptide bond. S-nitrosylation is another vital posttranslational modification which modulates protein stability. Caspases residing in the outer mitochondrial membrane are present as S-nitrothiols. On apoptotic signalling, they translocate to the cytosol, become denitrosylated and are activated in the cytosol. Post translational modifications like methylation, acetylation are also involved in diseased states.

We thus understand that, gene expression can be controlled intricately by targeting either the transcription machinery or the translation machinery in a spatio temporal manner. Controls exercised at post transcription and post translation levels have vital implications during disease signalling. This takes us to an interesting logical question on whether apart from the controls exercise at theses states can be totally shut off or activate a gene to initiate favorable responses inside the cell.

Genetic Switches

This is the right time to discuss an interesting experiment designed by Andr'e Lwoff, Francis Jacob and Jacques Monod at the Pasteur institute in Paris nearly 5 decades back. Their experiment showed that a strain of the bacterium E.coli irradiated with ultraviolet light halted their growth and after nearly 90 minutes lyses releasing a crop of viruses into the culture medium. These viruses (originally called lambda) are also called bacteriophages. In many bacteria the lambda virus is dormant but several other bacteria infected by the virus lyse, producing new phage. The normal growth and division when repeated produces a crop of new phages. This experiment clearly demonstrated that the virus switches between two states from the dormant state in the dividing bacterium to the activated state in the bacterium irradiated with UV light. This is a striking example of turning 'ON' or 'OFF' specific genes. We know that genes are the functional components of a living cell. Such living cells (bacterial or human) utilise only a subset of their genes to signal the production of other molecules. Therefore those genes which are expressed are termed 'ON' and those not expressed are turned 'OFF'. In other words we call this phenomenon regulation of gene expression.

This regulation of gene expression as an event takes place not only during the developmental cycle but throughout the life time of a diffrentiative cell.

Gene regulatory proteins and the specific sequences of DNA recognised by these proteins form the basic components of 'genetic switches'. It has been shown that nearly 80% of genetic material of gene switches alters the function of a gene. These genetic switches could be activated or deactivated by external signals, toxins, medium deprivation, stress etc. For example, continuous exposure to sunlight changes the colour of skin cells. Sunlight does not change the structure of the pigmentation gene but alters its function by turning the gene 'ON'. The brilliant discovery of gene switching ON and OFF in lambda phage revolutionised several new experiments in the field and established that the switching mechanism observed in E.coli and lambda phage also applied to eukaryotic cells.

Typical Gene Switching Systems in Cells

Skeletal muscle is composed of two types of fibers- slow twitch fibers which are innately vascular and fast twitch fibres which are deficient in blood vessels. In one of the clinical complications called Critical limb ischemia, blood flow to skeletal muscle is blocked leading to muscle wasting and eventually to the amputation of the limbs. Experiments have established that the genetic switch estrogen related receptor gamma (ERR gamma) when expressed in fast twitch fibers converts them into slow twitch fibers resulting in a significant increase in blood supply to the skeletal muscle. This is a classic case of treating a clinical complication without a pharmacological intervention.

TORC2 is a protein which promotes gluconeogenesis in liver under hypoglycemic conditions by functioning as a metabolic switch. This protein resides outside the nucleus under normal conditions but upon oxidative stress or starvation, shuttles to nucleus and activates a network of genes, vital for handling the insult. Mutations in the TORC protein have been shown to reduce the life expectancy. TORC mutated flies lose their lipid storing capability and have been shown to regain starvation and stress resistance when TORC is expressed in the nervous system.

Experiments on hypoxia tolerant fruit flies have shown that hairy, a transcriptional suppressor is critical for cell survival under hypoxic conditions. Hairy gene may shut off or hinder activation of many genes. On activation it restrains various signalling pathways allowing the cells to circumvent hypoxia. It activates a sort of cutback mechanism in cells culminating in energy conservation which is then used for important functions. Let us now, discuss genetic switches in bacteria.

Bacterial Response to a Single Signal- The Tryptophan Repressor as a Bacterial Switch

The genome of E.coli encodes approximately 4,200 proteins. Its chromosome comprises 4.6×10^6 bp. The expression of many of the genes in E.coli is heavily dependent on the availability of nutrients in the environment.

Tryptophan is a rare amino acid and is a precursor for niacin in eukaryotes. In bacteria, indole is formed from tryptophan. In plants it acts as precursor for biosynthesis the plant hormone auxins. Tryptophan is synthesized from chorismate in five steps catalyzed by three different enzymes which are produced by 5 genes as in Figure. The five genes include TrpE, TrpD, TrpC, TrpB, TrpA. Upstream to TrpE lies TrpL, operator, promoter and far from this stretch lies TrpR. These genes are arranged adjacent to each other on the chromosome as a single operon. The five genes are transcribed as a single mRNA molecule from a single promoter. When tryptophan in the growth medium enters the cell, the cell does not require these enzymes and therefore shuts off the production of these enzymes. The molecular mechanism of the tryptophan switch has been clearly understood and established. The tryptophan repressor is a member of the Helix-Turn-Helix family in which the promoter and the operator are arranged to facilitate classic switching.

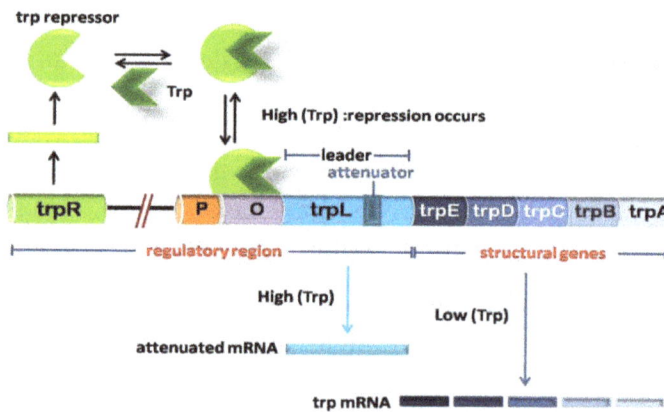

Gene regulation in Trp operon. When the level of trptophan
is high: Repression occurs in the circuit. Lower levels of tryptophan results in expression of a gene

In the first step, chorismate is converted to anthranilate. Anthranilate is converted into phosphoribosylanthranilate which is then converted into carboxy phenyl aminodeoxy ribulose -5-phosphate. After this indole- 3 –glycerol phosphate is formed which is ultimately converted into L-tryptophan via indole formation.

TrpL is a leader sequence and contains a critical region called TrpA meant for attenuation. TrpR produces inactive repressor called apo repressor which can't bind the operator tryptophan acts as corepresssor and activates apo repressor for shutting down genes. The repressor and apo repressor are dimers made up of identical Helix-Turn-Helix monomers. Tryptophan binding prepares the repressor for precise interaction with operator sequences. Trp repressor weakly regulates its own synthesis by binding with an operator site located in its promoter.

In a tryptophan switch, gene expression is regulated through a novel but simple mechanism. In order to bind to the operator DNA, the repressor protein should bind to the amino acid tryptophan through two of its molecules. The binding of tryptophan realigns the Helix-Turn-Helix motif of the repressor presenting it to the major groove of the DNA. When tryptophan is not present, the motif swings inward preventing the binding of the protein to the operator. The tryptophan repressor and the operator thus form an elegant switching device that turns gene ON and OFF.

Tryptophan repressor is an example of a negative repressible operon. (Negative with reference to repressor and repressible with reference to tryptophan). If the tryptophan is meager, bacterial structural genes necessary for converting chorismate to tryptophan are activated as the repressor made by TrpR is inactive and is unable to bind to the operator.

In the presence of tryptophan, structural genes need not transcribe as this may result in energy wastage. Tryptophan binds the inactive repressor making it functional.The active repressor binds to operator providing hindrance for RNA polymerase binding. This results in gene repression.

Attenuation

Premature termination of primary transcript in the leader region i.e. before the first structural genes is called attenuation. Attenuation is carried out by attenuator, a sequence within leader region of the tryptophan operon. At this site, choice is made by RNA polymerase either to terminate or continue transcription. Mutants with small deletions in this region produce tryptophan synthesizing enzymes even in the presence of tryptophan.

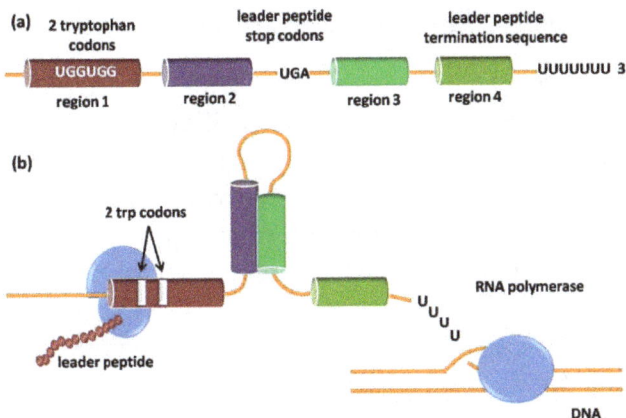

Termination of Transcription regulated by attenuation
(a) Stem-loop structures of the trp operon in the mRNA;
(b) Low level of trp full length mRNA made;
(c) High level transcription of the trp operon is prematurely halted

The Concept of the lac operon- Transcription Activation and Repression Control

Our earlier discussion on genetic switches gave us an idea of how genetic switches function in the cell and how they elegantly control cellular functions. We discussed the specific case of the simple tryptophan repressor which functions as a switch in bacterial systems. Let us now shift our discussions to complicated types of switching circuits in nature. The complexities in switching in such circuits are brought about by positive and negative controls in the circuit. The classic example of such a switching is the lac operon. The lac operon in E.coli is regulated by the lac repressor and CAP (Catabolite Activator Protein) and is controlled both negatively and positively at the transcriptional state.

Lactose Operon

The lac operon consists of a cluster of functionally related genes controlled by a single promoter. This operon includes the promoter and operator apart from three structural genes lacZ, lacy and lacA. LacZ codes for beta galactosidase, lacY codes for permease and lacA codes for transacetylase. Figure (a) explains the gene regulation in Lac operon. Here Beta galactosidase is a cytoplasmic protein which hydrolyses lactose, Permease acts as lactose importer, and transacetylase detoxifies toxic beta galactosides.

Lactose is a disaccharide which upon hydrolysis forms glucose and galactose in the cell and eventually to break it down while the CAP facilitates bacteria to utilise lactose in the absence of glucose. In the absence of lactose, it is not required for CAP to induce the expression of the lac operon. Here the lac repressor ensures that the lac operon is shut off. This facilitates the controlled region of lac operon to integrate two different signals, ensuring that the operon is expressed only when both conditions are met- lactose present and glucose absent. The other three possible combinations of signals keep the cluster of genes in the OFF state. The lac operon codes for proteins that transport lactose into small amount of permease is found even under repressive condition. The regulator produces a repressor which binds to operator and prevents RNA polymerase from binding to promotor and thus inhibits transcription of three structural genes. The gene regulation process in lac operon is illustrated in Figure (b).

(a)

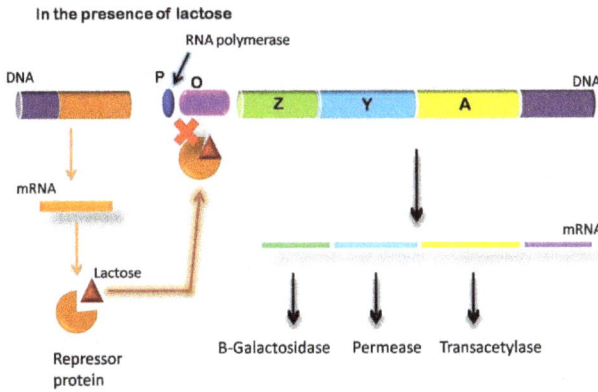

(b)

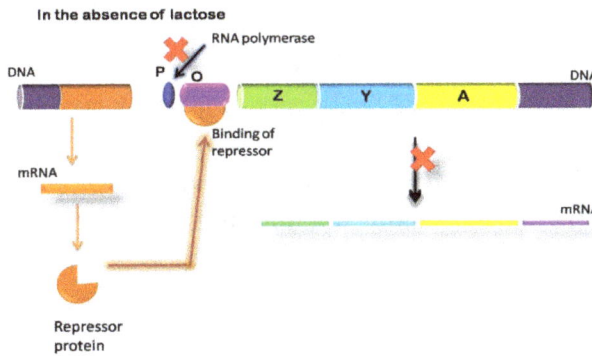

Gene regulation through the Lac operon (a) In the presence of Lactose;
(b) In the absence of Lactose

When both Glucose and Lactose are Absent

Since the lactose (inducer) is not available to bind to the repressor protein, hence the repressor binds to the promoter region and terminates the process of transcription. As a result there is no gene expression.

When both Glucose and Lactose are Present

Under a such condition, the bacteria prefer glucose and utilise lactose only when glucose is exhausted, thus recording two growth curves. This is called diauxic growth.

When Lactose is Present and Glucose is Absent

Lactose is taken in with the help of permease and is converted into allo-lactose. Allo-lactose binds to the repressor and makes it non functional and thus the three structural genes are transcribed.

Thus the lac– operon is an example of a negative inducible operon -negative with reference to effect of repressor on transcription of structural genes and inducible with reference to effect of lactose on structural gene transcription. Under this situation the levels of cAMP are high.

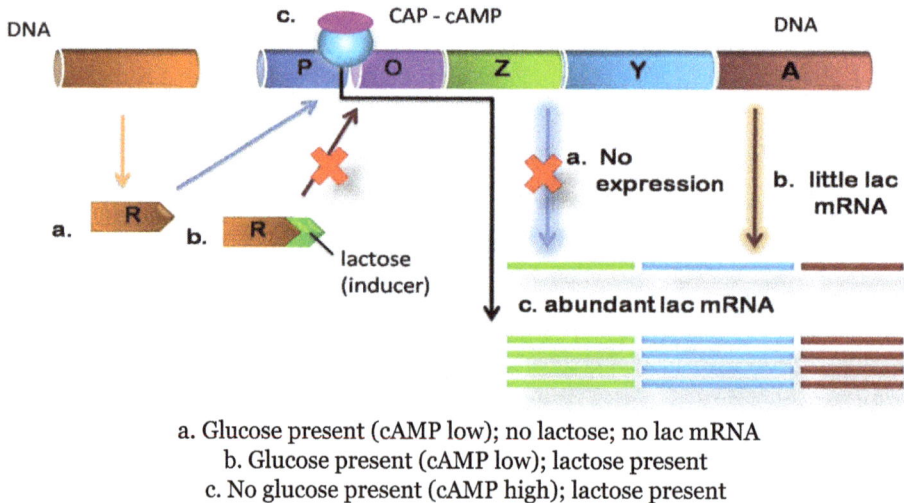

a. Glucose present (cAMP low); no lactose; no lac mRNA
b. Glucose present (cAMP low); lactose present
c. No glucose present (cAMP high); lactose present

Riboswitches- Regulatory Functions of RNA

The central dogma of life has always portrayed the nucleic acids as the blue print for a cell assigning the regulatory and enzymatic functions to the proteins sysnthesised in the cell. Recent work on RNA over the past decade and explorations of its newer functions in the cell strongly challenge the text book view of the central dogma. Though RNA molecules have been shown to be involved in cleavage, splicing and translation and novel gene regulatory mechanisms operating at both the DNA and mRNA level have been explored in detail. Newer functions of specific RNAs that can function as sensors of vitamin B1, B2 and B12 cofactors, have taken center stage. Riboswitches are a fascinating type of RNA structures that regulate gene expression both at the transcription and translation levels by binding to small molecules (ligands). These are structures that form in a messenger RNA and are involved predominantly in gene regulation events in bacteria.

These riboswitches regulate gene expression through the formation of alternative structures which either prematurely terminate transcription or inhibit the initiation of translation when they are in the repressing conformation. Riboswitches regulate the synthesis of Riboflavin, Thiamin and Cobalamin and the metabolism of Methionine, Lysine and Purines. Riboswitches are present in bacterial species, in fungi and in plants. More than 2% of the *Bacillus subtilis* genome has been shown to be regulated by riboswitches.

What do we know about Riboswitches?

Riboswitches fold into compact RNA secondary structures which comprise a base stem, a central multi loop and several branching hair pins as shown in Figure.

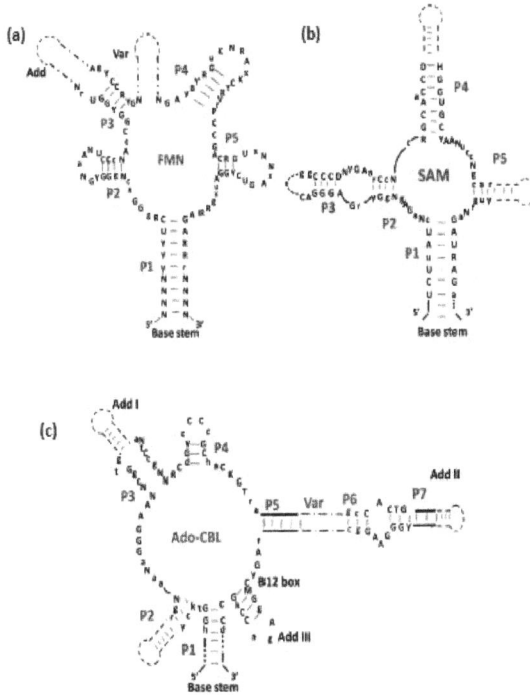

Structure of riboswitches; (a) RFN-element; (b) G-box; (c) B12-element

The riboswitches distinguish themselves strikingly form other regulatory systems through two features. The first deals with the fact that riboswitches are present across diverse organisms. For example the THI elements are observed in eubacteria, archea and eukaryotes. The S-boxes, G-boxes and L-boxes are observed in gram-positive bacteria from the *Bacillus, Thermotogale* and *Bacteroidetes* species. The next outstanding feature of the riboswitches is that they regulate diverse processes as riboflavin and transport, thiamin synthesis and transport, purine metabolism and synthesis etc as in Table.

Table	Riboswitches and their properties	
Riboswitches	Functional system	Ligand
RFN-element	Riboflavin biosynthesis and transport	FMN (flavin mononucleotide)
THI-element	Thiamin biosynthesis ; transport of thiamin and related compounds	TPP (thiamin pyrophosphate)
B12-element	Cobalamin biosynthesis; transport of cobalamin and related compounds; cobalt transport; cobalamin-independent isoenzymes of cobalamin- dependent enzymes	Coenzyme B12 (adenosylcobalamin)
S-box	Methionine biosynthesis and transport SAM metabolism	SAM (S- adenosylmethionine)

Mechanism of Function of Riboswitches

As stated earlier, regulation through riboswitches involves the formation of alternative structures.

Condition 1: Repressed State

During repression two RNA structures are formed, the small molecule ligand binds to the structure and stabilizes the switch forming the regulatory hairpin. This hairpin sequesters to the ribosome binding site and can function either to terminate the transcription process or to inhibit the initiation of translation.

Condition 2: Derepressed State

During the derepressed state, the riboswitches is not involved in ligand binding and therefore forms an alternative structure comprising the complementary regions in the riboswitches base stems and a portion of the regulatory hairpin. When the riboswitches directly sequesters the site of translation initiation variations occur in this switch. On change of parity the riboswitches functions as an alternative to the regulatory hairpin, activating gene expression in the presence of the ligand and repressing gene expression when it is not bound.

How Relevant is the Study of Genetic Switches to Systems Biology?

Molecular Biology techniques and precise genetics over the past decades have established the genetic switch as a leading theme of gene regulation. The availability of vast amounts of data from genomics, proteomics and high throughput experimentation, have enabled biology to move from the component centric paradigm to a systems level quest to understand how specific parts of system function together to carry out complex functions. Systems Biology approaches would help illustrate the factors that regulate the efficiency of the switch. Systems Biology can help in modelling the long range interactions between the regulatory proteins in the network and the cooperatively involved in such tight negative auto regulatory networks. The evolution of the switch is an interesting paradigm to be explored using Systems Biology. A question like how robust is the switch against molecular level alterations, how stochastic is the gene regulatory process under such switching would throw interesting discussions on these themes. The construction of genetic toggle switches in vivo would also be an off shoot of such explorations.

References

- Bird A (2002). "DNA methylation patterns and epigenetic memory". Genes Dev. 16 (1): 6–21. doi:10.1101/gad.947102. PMID 11782440
- Whiteside, ST; Goodbourn, S (April 1993). "Signal transduction and nuclear targeting: regulation of transcription factor activity by subcellular localisation.". Journal of Cell Science. 104 (4): 949–55. PMID 8314906

- Bruce Alberts; Alexander Johnson; Julian Lewis; Martin Raff; Keith Roberts; Peter Walter (2007). Molecular Biology of the Cell (Fifth ed.). Garland Science. pp. 1268 pages. ISBN 0-8153-4105-9

- Jacob, F.; Monod, J. (1961). "Genetic regulatory mechanisms in the synthesis of proteins". J. Mol. Biol. 3: 318–356. doi:10.1016/s0022-2836(61)80072-7. PMID 13718526

- Vihervaara, A; Sistonen, L (15 January 2014). "HSF1 at a glance.". Journal of Cell Science. 127 (Pt 2): 261–6. doi:10.1242/jcs.132605. PMID 24421309

- Weaver, Robert J. (2007). "Part V: Post-transcriptional events". Molecular Biology. Boston: McGraw Hill Higher Education. ISBN 0-07-110216-7

- Struhl, K (9 July 1999). "Fundamentally different logic of gene regulation in eukaryotes and prokaryotes.". Cell. 98 (1): 1–4. doi:10.1016/s0092-8674(00)80599-1. PMID 10412974

- Napolitano, G; Lania, L; Majello, B (May 2014). "RNA polymerase II CTD modifications: how many tales from a single tail.". Journal of cellular physiology. 229 (5): 538–44. doi:10.1002/jcp.24483. PMID 24122273

- Mims C, Nash A, Stephen J. Mims' pathogenesis of infectious disease. 2001. 5th. Academic press. ISBN 0-12-498264-6

- Calo, E; Wysocka, J (7 March 2013). "Modification of enhancer chromatin: what, how, and why?". Molecular Cell. 49 (5): 825–37. doi:10.1016/j.molcel.2013.01.038. PMC 3857148. PMID 23473601

- Vogelstein B, Papadopoulos N, Velculescu VE, Zhou S, Diaz LA, Kinzler KW (2013). "Cancer genome landscapes". Science. 339 (6127): 1546–58. doi:10.1126/science.1235122. PMC 3749880. PMID 23539594

- Farazi TA, Spitzer JI, Morozov P, Tuschl T (2011). "miRNAs in human cancer". J. Pathol. 223 (2): 102–15. doi:10.1002/path.2806. PMC 3069496. PMID 21125669

Fluctuations in Gene Expression

DNA sequences that are close together can be inherited together. This process is known as genetic linkage. The genetic markers found together have less of a tendency of being separated. The topics discussed in the chapter are of great importance to broaden the existing knowledge on systems biology.

Genetic Linkage

Genetic linkage is the tendency of DNA sequences that are close together on a chromosome to be inherited together during the meiosis phase of sexual reproduction. Two genetic markers that are physically near to each other are unlikely to be separated onto different chromatids during chromosomal crossover, and are therefore said to be more *linked* than markers that are far apart. In other words, the nearer two genes are on a chromosome, the lower the chance of recombination between them, and the more likely they are to be inherited together. Markers on different chromosomes are perfectly *unlinked*.

Genetic linkage is the most prominent exception to Gregor Mendel's Law of Independent Assortment. The first experiment to demonstrate linkage was carried out in 1905. At the time, the reason why certain traits tend to be inherited together was unknown. Later work revealed that genes are physical structures related by physical distance.

The typical unit of genetic linkage is the centimorgan (cM). A distance of 1 cM between two markers means that the markers are separated to different chromosomes on average once per 100 meioses.

Discovery

Gregor Mendel's Law of Independent Assortment states that every trait is inherited independently of every other trait. But shortly after Mendel's work was rediscovered, exceptions to this rule were found. In 1905, the British geneticists William Bateson, Edith Rebecca Saunders and Reginald Punnett, cross-bred pea plants in experiments similar to Mendel's. They were interested in trait inheritance in the sweet pea and were studying two genes—the gene for flower colour (*P*, purple, and *p*, red) and the gene affecting the shape of pollen grains (*L*, long, and *l*, round). They crossed the pure lines *PPLL* and *ppll* and then self-crossed the resulting *PpLl* lines. According to Mendelian genetics, the expected phenotypes would occur in a 9:3:3:1 ratio of PL:Pl:pL:pl. To their surprise, they observed an increased frequency of PL and pl and a decreased frequency of Pl and pL:

Bateson, Saunders, and Punnett experiment		
Phenotype and genotype	Observed	Expected from 9:3:3:1 ratio
Purple, long (*P_L_*)	284	216
Purple, round (*P_ll*)	21	72
Red, long (*ppL_*)	21	72
Red, round (*ppll*)	55	24

Their experiment revealed linkage between the *P* and *L* alleles and the *p* and *l* alleles. The frequency of *P* occurring together with *L* and *p* occurring together with *l* is greater than that of the recombinant *Pl* and *pL*. The recombination frequency is more difficult to compute in an F2 cross than a backcross, but the lack of fit between observed and expected numbers of progeny in the above table indicate it is less than 50%.

The understanding of linkage was expanded by the work of Thomas Hunt Morgan. Morgan's observation that the amount of crossing over between linked genes differs led to the idea that crossover frequency might indicate the distance separating genes on the chromosome. The centimorgan, which expresses the frequency of crossing over, is named in his honour.

Linkage Map

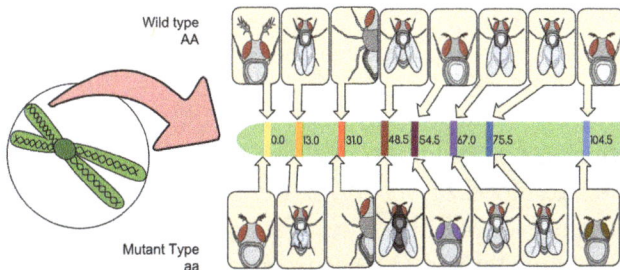

Thomas Hunt Morgan's *Drosophila melanogaster* genetic linkage map.

This was the first successful gene mapping work and provides important evidence for the chromosome theory of inheritance. The map shows the relative positions of alleles on the second Drosophila chromosome. The distances between the genes (centimorgans) are equal to the percentages of chromosomal crossover events that occur between different alleles.

A linkage map (also known as a genetic map) is a table for a species or experimental population that shows the position of its known genes or genetic markers relative to each other in terms of recombination frequency, rather than a specific physical distance along each chromosome. Linkage maps were first developed by Alfred Sturtevant, a student of Thomas Hunt Morgan.

A genetic map is a map based on the frequencies of recombination between markers during crossover of homologous chromosomes. The greater the frequency of recombination (segregation) between two genetic markers, the further apart they are assumed to be. Conversely, the lower the frequency of recombination between the markers, the smaller the physical distance between them. Historically, the markers originally used were detectable phenotypes (enzyme production, eye colour) derived from coding DNA sequences; eventually, confirmed or assumed noncoding DNA sequences such as microsatellites or those generating restriction fragment length polymorphisms (RFLPs) have been used.

Genetic maps help researchers to locate other markers, such as other genes by testing for genetic linkage of the already known markers.

A genetic map is not a physical map (such as a radiation reduced hybrid map) or gene map.

Linkage Analysis

Linkage analysis may be either parametric (if we know the relationship between phenotypic and genetic similarity) or non-parametric. Parametric linkage analysis is the traditional approach, whereby the probability that a gene important for a disease is linked to a genetic marker is studied through the LOD score, which assesses the probability that a given pedigree, where the disease and the marker are cosegregating, is due to the existence of linkage (with a given linkage value) or to chance. Non-parametric linkage analysis, in turn, studies the probability of an allele being identical by descent with itself.

Parametric Linkage Analysis

The LOD score (logarithm (base 10) of odds), developed by Newton Morton, is a statistical test often used for linkage analysis in human, animal, and plant populations. The LOD score compares the likelihood of obtaining the test data if the two loci are indeed linked, to the likelihood of observing the same data purely by chance. Positive LOD scores favour the presence of linkage, whereas negative LOD scores indicate that linkage is less likely. Computerised LOD score analysis is a simple way to analyse complex family pedigrees in order to determine the linkage between Mendelian traits (or between a trait and a marker, or two markers).

The method is described in greater detail by Strachan and Read. Briefly, it works as follows:

1. Establish a pedigree

2. Make a number of estimates of recombination frequency

3. Calculate a LOD score for each estimate

4. The estimate with the highest LOD score will be considered the best estimate

The LOD score is calculated as follows:

$$LOD = Z = \log_{10} \frac{\text{probability of birth sequence with a given linkage value}}{\text{probability of birth sequence with no linkage}} = \log_{10} \frac{(1-\theta)^{NR} \times \theta^{R}}{0.5^{(NR\ R)}}$$

NR denotes the number of non-recombinant offspring, and R denotes the number of recombinant offspring. The reason 0.5 is used in the denominator is that any alleles that are completely unlinked (e.g. alleles on separate chromosomes) have a 50% chance of recombination, due to independent assortment. 'θ' is the recombinant fraction, i.e. the fraction of births in which recombination has happened between the studied genetic marker and the putative gene associated with the disease. Thus, it is equal to R / (NR + R)

By convention, a LOD score greater than 3.0 is considered evidence for linkage, as it indicates 1000 to 1 odds that the linkage being observed did not occur by chance. On the other hand, a LOD score less than -2.0 is considered evidence to exclude linkage. Although it is very unlikely that a LOD score of 3 would be obtained from a single pedigree, the mathematical properties of the test allow data from a number of pedigrees to be combined by summing their LOD scores. A LOD score of 3 translates to a p-value of approximately 0.05, and no multiple testing correction (e.g. Bonferroni correction) is required.

Recombination Frequency

Recombination frequency is a measure of genetic linkage and is used in the creation of a genetic linkage map. Recombination frequency (θ) is the frequency with which a single chromosomal crossover will take place between two genes during meiosis. A centimorgan (cM) is a unit that describes a recombination frequency of 1%. In this way we can measure the genetic distance between two loci, based upon their recombination frequency. This is a good estimate of the real distance. Double crossovers would turn into no recombination. In this case we cannot tell if crossovers took place. If the loci we're analysing are very close (less than 7 cM) a double crossover is very unlikely. When distances become higher, the likelihood of a double crossover increases. As the likelihood of a double crossover increases we systematically underestimate the genetic distance between two loci.

During meiosis, chromosomes assort randomly into gametes, such that the segregation of alleles of one gene is independent of alleles of another gene. This is stated in Mendel's Second Law and is known as the law of independent assortment. The law of independent assortment always holds true for genes that are located on different chromosomes, but for genes that are on the same chromosome, it does not always hold true.

As an example of independent assortment, consider the crossing of the pure-bred homozygote parental strain with genotype *AABB* with a different pure-bred strain with genotype *aabb*. A and a and B and b represent the alleles of genes A and B. Crossing these homozygous parental strains will result in F1 generation offspring that are double heterozygotes with genotype AaBb. The F1 offspring AaBb produces gametes that are *AB*, *Ab*, *aB*, and *ab* with equal frequencies (25%) because the alleles of gene A assort independently of the alleles for gene B during meiosis. Note that 2 of the 4 gametes (50%)—*Ab* and *aB*—were not present in the parental generation. These gametes represent recombinant gametes. Recombinant gametes are those gametes that differ from both of the haploid gametes that made up the original diploid cell. In this example, the recombination frequency is 50% since 2 of the 4 gametes were recombinant gametes.

The recombination frequency will be 50% when two genes are located on different chromosomes or when they are widely separated on the same chromosome. This is a consequence of independent assortment.

When two genes are close together on the same chromosome, they do not assort independently and are said to be linked. Whereas genes located on different chromosomes assort independently and have a recombination frequency of 50%, linked genes have a recombination frequency that is less than 50%.

As an example of linkage, consider the classic experiment by William Bateson and Reginald Punnett. They were interested in trait inheritance in the sweet pea and were studying two genes—the gene for flower colour (*P*, purple, and *p*, red) and the gene affecting the shape of pollen grains (*L*, long, and *l*, round). They crossed the pure lines *PPLL* and *ppll* and then self-crossed the resulting *PpLl* lines. According to Mendelian genetics, the expected phenotypes would occur in a 9:3:3:1 ratio of PL:Pl:pL:pl. To their surprise, they observed an increased frequency of PL and pl and a decreased frequency of Pl and pL.

Bateson and Punnett experiment		
Phenotype and genotype	Observed	Expected from 9:3:3:1 ratio
Purple, long (*P_L_*)	284	216
Purple, round (*P_ll*)	21	72
Red, long (*ppL_*)	21	72
Red, round (*ppll*)	55	24

Their experiment revealed linkage between the *P* and *L* alleles and the *p* and *l* alleles. The frequency of *P* occurring together with *L* and with *p* occurring together with *l* is greater than that of the recombinant *Pl* and *pL*. The recombination frequency is more difficult to compute in an F2 cross than a backcross, but the lack of fit between observed and expected numbers of progeny in the above table indicate it is less than 50%.

The progeny in this case received two dominant alleles linked on one chromosome (referred to as coupling or cis arrangement). However, after crossover, some progeny could have received one parental chromosome with a dominant allele for one trait (e.g. Purple) linked to a recessive allele for a second trait (e.g. round) with the opposite being true for the other parental chromosome (e.g. red and Long). This is referred to as repulsion or a trans arrangement. The phenotype here would still be purple and long but a test cross of this individual with the recessive parent would produce progeny with much greater proportion of the two crossover phenotypes. While such a problem may not seem likely from this example, unfavourable repulsion linkages do appear when breeding for disease resistance in some crops.

The two possible arrangements, cis and trans, of alleles in a double heterozygote are referred to as gametic phases, and *phasing* is the process of determining which of the two is present in a given individual.

When two genes are located on the same chromosome, the chance of a crossover producing recombination between the genes is related to the distance between the two genes. Thus, the use of recombination frequencies has been used to develop linkage maps or genetic maps.

However, it is important to note that recombination frequency tends to underestimate the distance between two linked genes. This is because as the two genes are located farther apart, the chance of double or even number of crossovers between them also increases. Double or even number of crossovers between the two genes results in them being cosegregated to the same gamete, yielding a parental progeny instead of the expected recombinant progeny. As mentioned above, the Kosambi and Haldane transformations attempt to correct for multiple crossovers.

Variation of Recombination Frequency

While recombination of chromosomes is an essential process during meiosis, there is a large range of frequency of cross overs across organisms and within species. Sexually dimorphic rates of recombination are termed heterochiasmy, and are observed more often than a common rate between male and females. In mammals, females often have a higher rate of recombination compared to males. It is theorised that there are unique selections acting or meiotic drivers which influence the difference in rates. The difference in rates may also reflect the vastly different environments and conditions of meiosis in oogenesis and spermatogenesis.

Meiosis Indicators

With very large pedigrees or with very dense genetic marker data, such as from whole-genome sequencing, it is possible to precisely locate recombinations. With this type of genetic analysis, a meiosis indicator is assigned to each position of the genome for each meiosis in a pedigree. The indicator indicates which copy of the parental chromosome contributes to the transmitted gamete at that position. For example, if the allele from the 'first' copy of the parental chromosome is transmitted, a '0' might be assigned to that meiosis. If the allele from the 'second' copy of the parental chromosome is transmitted, a '1' would be assigned to that meiosis. The two alleles in the parent came, one each, from two grandparents. These indicators are then used to determine identical-by-descent (IBD) states or inheritance states, which are in turn used to identify genes responsible for diseases.

Structure ad Genetic Map of Lambda Phage

Viruses are obligate intracellular parasites. Bacteriophages are viruses infecting bacteria. They multiply inside the host system through partial or complete utilization of the host biosynthetic machinery.

Bacteriophages may be

- RNA phages such as Q-beta

- Filamentous with single stranded DNA such as M13

- T- even phages including T2,T4,T6 infecting *E.coli*

- Temperate phages like lambda and mu

- Spherical phages with single stranded DNA like PhiX174

The genetic material in bacteriophage may either be RNA or DNA but not both. The nucleic acids of phages contain unusual or modified bases which enable to circumvent the degradation of the host nuclease. Further, the number of genes in bacteriophage varies with the complexity of the genome. Complex phages have more than 100 genes while simple phages have only 3-5 genes.

Structure of the Lambda Phage

Lambda phage is a *temperate phage*. The genes of the lambda phage make a single DNA molecule-the chromosome wrapped within a protein coat, composed of 12-15 different proteins all of which are encoded by the lambda chromosome. The coat is structurally composed of an icosahedral head with a diameter of 64nm and a tail, 150 nm in length as shown in Figure. The *head* is composed of double stranded linear DNA surrounded by a capsid made up of protein capsomers. At the 5' end of each strand are 12 nucleotide long sequences complementary to each other. Thus on circularization, the bacteriophage DNA has 48,514 base pairs. The 5'ends are called cos sites and the opposite of cos site is att site meant for attachment. The lambda phage attaches to the cell surface of *E.coli* through its tail, making a hole in the cell wall. It thus pushes its chromosome into the bacterium *E.coli,* leaving behind the protein coat.

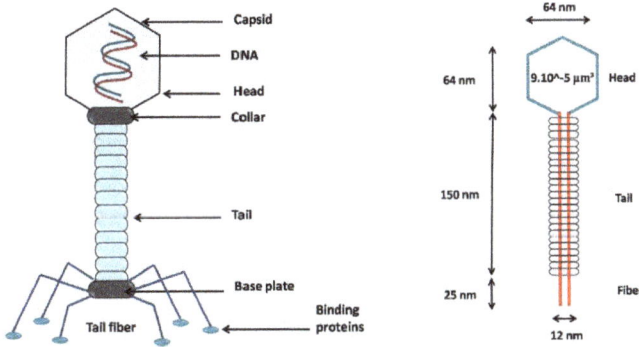

(a) Structure of a Lambda phage , (b) measurement of different parts of a Lambda phage

First stage of infection involves a process called *adsorption*. Adsorption involves landing and attachment. Tail fibres play a critical role in this stage. Tail less phages use analogous structures for adsorption. Specific receptors on the bacterial cell like proteins, lipopolysaccharides, pili apart from lipoproteins are exploited by phages for attachment. This is reversible condition. Base plate components mediate permanent binding.

Second stage in infection process is *penetration*. The sheath of phages contracts resulting in insertion of hollow tail fiber through bacterial envelope. Some phages utilize their enzymes to digest components of bacterial envelope. Nucleic acid is inserted inside bacterial cell via hollow tail. Remaining part of the phage outside bacteria is called ghost. Thus in nutshell penetration involves contraction of sheath till DNA insertion.

Some phages upon infecting bacteria lyse the bacterial cell after forming their progeny. Such phages are called virulent. Some phages integrate their genome into bacterial genome and can remain inside host without harming them but under drastic conditions can become virulent and can causes host cell lysis. Such phages which normally follow lysogeny but under drastic conditions become lytic are called *temperate phages*.

Receptor Targeting λ- Phage

The λ - phage uses the maltose pore LamB for delivering its genetic material into the host cell. The phage binds to the cells of the target *E.coli* and the J-protein in the tip of its tail interacts with the LamB, gene product of *E.coli* (LamB is a porin molecule and is a part of the maltose operon). Most of the *E.coli* K-12 mutations resistant to λ phage are located in two genetic regions *malA* and *malB*. LamB is composed of three identical subunits, each of which is formed by an 18-stranded antiparallel β-barrel, which forms a wide channel with a diameter of about 2.5 nm. The phage consists of a hollow tube composed of 32 −stacked discs, each of which has a 3nm central hole to eject its genetic material into the host. Lambda phage uses this channel for ejecting its genetic material. After injecting the DNA into bacteria, the double stranded linear DNA circularizes due to the presence of cos sites and site specific nucleases cut DNA at the att site of the phage DNA.

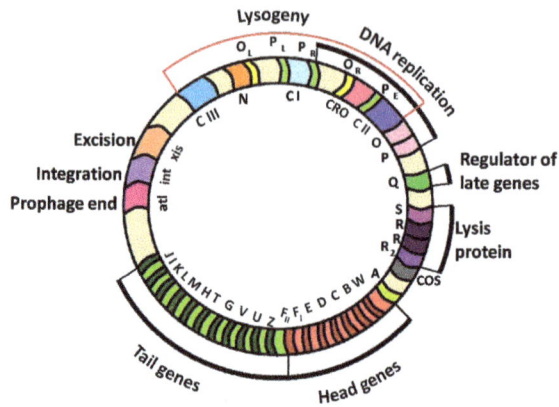

Genetic mapping of lambda phage

Table: The genes involved in the switching mechanism of the lambda phage	
Name of gene/promotor/operator	Function
Att	Provides site of attachment for phage to host chromosome
cI	Repressor protein
cII	Coding for promotor establishment activator protein
cIII	Codes for stabilizing protein
OR	Operator right
PR	Promotor right
OL	Operator left
PL	Promotor left
Cro	Gene for second repressor
N	Positive regulator counteracting rho dependent termination
J through U	Genes encoding tail proteins
Z through A	Genes encoding head proteins
Int	Gene encoding integrase
Xis	Encodes excisionase
O and P	Encode proteins involved in replication of Lambda DNA
Q	Encodes anti terminator protein

Bacteriophage lambda is *episomic* and consequently its genome exists in at least two states within which genetic recombination is possible. This allows the construction of two genetic maps termed vegetative and prophage after these states. In the vegetative state, the replication of the lambda genome is independent of the replication of the host genome. Such replica are finally packaged into the head of the mature phage as single duplex DNA molecules, 15 to 17 microns in length. These molecules contain ~ 47,000 base pairs, to accommodate 40 to 45 structural genes. In the prophage state the viral genome integrates into the host genome replicating in synchrony with the host genome. Lambda genes are organized into operons. The Left operon genes are meant for recombination and integration resulting in lysogeny, while right operon and the late operon genes are meant for lysis. The genetic map of the lambda phage is shown in figure.

The lambda phage infects the bacterium directing it to two different fates. In some of the cells, as the infection happens the different set of phage genes are turned ON and OFF in a precisely regulated manner. The lambda chromosome is replicated, newer head and tail proteins are synthesized, forming new phage particles within the bacterium. As the phage chromosome begins to replicate, the phage gene cI, is expressed. The product of cI is the bacteriophage lambda repressor which keeps the other phage genes in the OFF state. When exposed to ultraviolet light, the inert phage genes (lytic genes) are switched ON and the repressor gene is switched OFF. Nearly 45 minutes after the infection the bacterial cell 'lyses' releasing around 100 new progeny phage.

In the other population of cells, the injected phage chromosome turns OFF all the phage genes except one. The single phage chromosome called the prophage now becomes a part of the host chromosome. The bacterium carrying the dormant phage chromosome is called the lysogen. As the lysogen grows, the prophage is passively replicated with the host genome and distributed to the progeny bacteria. Thus, we understand that the phage genes upon exposure of a lysogen to a signal such as UV, switch from their stable lysogenic state to a lytic growth state. The switch from the lysogeny to the lytic growth is termed induction.

Lytic and Lysogeny Decisions

Lysis

The lysis-lysogeny decision of the temperate lambda phage has emerged as a novel paradigm for understanding developmental genetic networks. *E.coli* and the lambda phage establish synergistic relationships. The lambda phage may exist in a dormant lysogenic state, passively replicating with the host chromosome or may fall into the lytic cycle generating progeny phages, killing their hosts. The lambda phage therefore makes a decision to follow either the lytic or the lysogenic pathway.

When the lambda phage follows the lytic pathway, it replicates its DNA autonomously, expresses a set of genes, and assembles the virions, resulting in lysis of the host. If the lysogenic state continues over a long time, a stable lysogen is established in the circuit and the prophage is integrated to the host genome. This turns OFF the expression of the lytic genes.

Inducing signals like UV light that damage the DNA, force the lambda phage to a SOS response and the lysogenic state switches irreversibly to the lytic phase as shown in Figure. The lambda phage thus behaves as a biphasic switch.

Growth and induction of the Lambda lysogen

The genetic map of the lambda phage is shown in Figure.

Genetic map of the lambda phage

The gene arrangement and sites involved in switching are shown in Figures.

The biphasic switch

Arrangement of genes and sites of OR

Function of the Switch in Lambda Phage

In order to understand how switching happens between the lysis and lysogeny states in the lambda phage, we focus on two regulatory genes CI and cro and a regulatory region O_R called the right operator as shown in Figure. During the lysogeny phase CI is switched ON and cro is OFF. The operator O_R is constituted of three binding sites O_{RI}, O_{RII} and O_{RIII} which overlap two promoters P_{RM} and P_R which oppose each other as shown in Figure. The promoter PR drives the transcription of lytic genes and P_{RM}, the transcription of the CI gene. During the lysogenic state, the lambda repressor at O_R is bound at O_{RI} and O_{RII}, sites adjacent to each other. At these sites, the repressor represses the right ward transcription from the promoter P_R. the expression of cro and other lytic genes is therefore turned OFF. At the same time, it also transcribes its own gene from the promoter P_{RM} as shown in Figure. When induced, the repressor leaves the operator and transcription from P_R is initiated spontaneously. Please note that P_R is a stronger promoter than PRM. As transcription begins, the CRO protein is made which binds first to O_{RIII} abolishing the synthesis of the repressor.

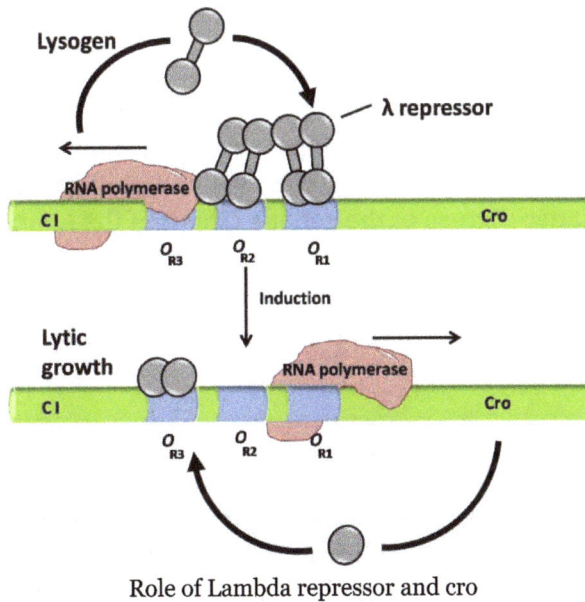

Role of Lambda repressor and cro

Lysogeneic State

We know that the lambda repressor is required to regulate the transcription of its own gene. We then question how this gene is switched ON to establish lysogeny during the viral infection. The repressor is transcribed initially from the promoter P_{RE} (Promoter for repressor establishment) as shown in Figure. This transcription is activated by CII, a product of another phage gene. Thus, a new repressor CI is made and it activates its own transcription from P_{RM}. This switches OFF the other phage genes including CII. Thus we see the establishment of lysogeny in lambda phage, even in the absence of the inducer signal.

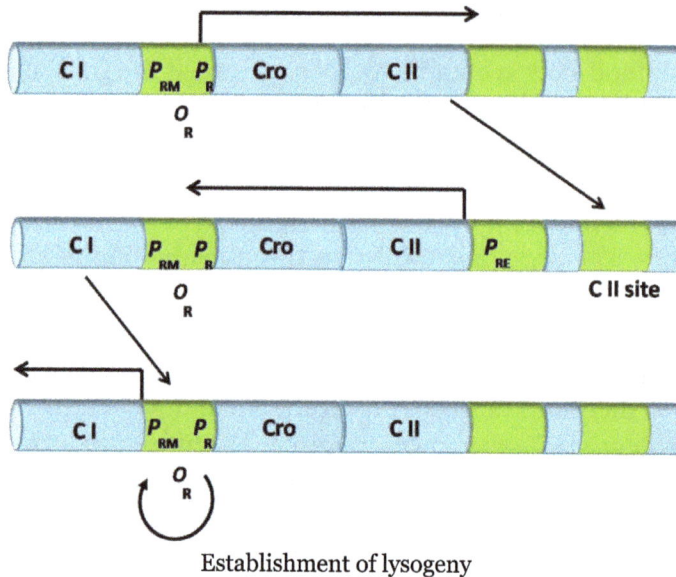

Establishment of lysogeny

The life styles of the phage lambda present a classic case of complex genetic control circuits. It is interesting to understand a small set of regulatory proteins yielding a complex set of temporally controlled macro molecular interactions in a simple organism like the lambda phage. Systems biology approaches will help understand the function of specific modules of these regulatory domains and will help understand the kinetic behavior and quantitative picture of the genetic circuit of the lambda phage.

Noise in Gene Expression

Fluctuations in Gene Expression

We now begin a very interesting discussion that clearly demonstrates the transition from biology to systems biology. We introduce here, the concept of 'noise' that is inherent in biological systems and discuss how systems are optimized to function under stochastic fluctuations. We would be using a number of new terms that quantify or statistically describe the system performance in various environments.

All biological systems are intrinsically heterogeneous. This heterogeneity may be brought in by the stochastic fluctuations of several processes involved in gene regulation either at the level of transcription or at the level of translation. Events like mRNA synthesis from DNA during transcription, synthesis of peptides from RNA during translation and mRNA degradation are highly stochastic. Genes in complex regulatory networks carry inherent expression noise. This creates isogenic cell populations which show variation in protein levels from cell to cell. Such random fluctuations in protein expression levels occur even in homogeneous environments of clonal cell populations.

Such fluctuations when quantified, describe the term 'biological noise' or 'gene expression noise'. Cell to cell variation in mRNA levels also contribute to expression noise.

Is noise an advantage to Biological Systems?

Noise is generally considered undesirable and unpredictable. Since parameters such as low reactant numbers, significantly influence the statistical fluctuations in the reaction rates as well as the number of the molecules involved in the reaction, living cells are perceived as biochemical reactors which show inherent noise characteristics. Since noise develops in the system during regulatory processes, organisms exploit this stochastic fluctuation to cause diversity in population. The best illustration of this diversity can be explained through lambda phage, a bacterial virus that infects populations of *E.coli*. The lambda phage undergoes a lytic cycle, where the lambda DNA is replicated several times, expresses the lysis proteins and the genes involved in the head and tail of the phage. During the lysogenic pathway, the phage DNA under different conditions integrates itself into the host cell chromosome. The lambda DNA now called a prophage, resides within the host genome without causing harm to the host.

Experimental evidences from the past decade have established that the most critical components of genetic circuits and gene regulatory networks show significant, unavoidable fluctuations in their expression levels and gene activity. A biological system should stabilize itself against such fluctuations in its gene regulatory cascades, since these cascades control cellular differentiation in developing embryos. Recent studies have shown that noise can provide critical functions that deterministic gene circuits cannot achieve.

Cellular Noise

Cellular noise is random variability in quantities arising in cellular biology. For example, cells which are genetically identical, even within the same tissue, are often observed to have different expression levels of proteins, different sizes and structures. These apparently random differences can have important biological and medical consequences.

Cellular noise was originally, and is still often, examined in the context of gene expression levels – either the concentration or copy number of the products of genes within and between cells. As gene expression levels are responsible for many fundamental properties in cellular biology, including cells' physical appearance, behaviour in response to stimuli, and ability to process information and control internal processes, the presence of noise in gene expression has profound implications for many processes in cellular biology.

Definitions

The most frequent quantitative definition of noise is the coefficient of variation:

$$\eta_X = \frac{\sigma_X}{\mu_X},$$

where η_X is the noise in a quantity X, μ_X is the mean value of X and σ_X is the standard deviation of X. This measure is dimensionless, allowing a relative comparison of the importance of noise, without necessitating knowledge of the absolute mean.

Another quantity often used for mathematical convenience is the Fano factor:

$$F_X = \frac{\sigma_X^2}{\mu_X}.$$

Intrinsic and Extrinsic Noise

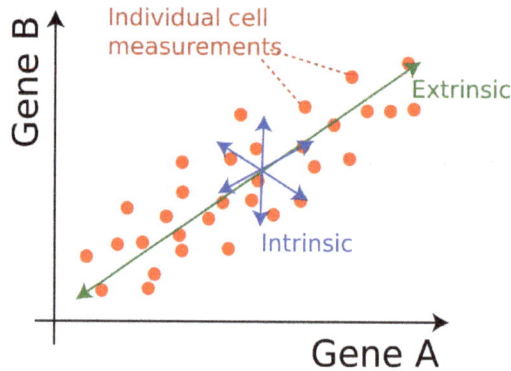

A schematic illustration of a dual reporter study. Each data point corresponds to a measurement of the expression level of two identically-regulated genes in a single cell: the scatter reflects measurements of a population of cells. Extrinsic noise is characterised by expression levels of both genes covarying between cells, intrinsic by internal differences.

Cellular noise is often investigated in the framework of *intrinsic* and *extrinsic* noise. Intrinsic noise refers to variation in identically-regulated quantities within a single cell: for example, the intra-cell variation in expression levels of two identically-controlled genes. Extrinsic noise refers to variation in identically-regulated quantities between different cells: for example, the cell-to-cell variation in expression of a given gene.

Intrinsic and extrinsic noise levels are often compared in dual reporter studies, in which the expression levels of two identically-regulated genes (often fluorescent reporters like GFP and YFP) are plotted for each cell in a population.

Sources

Note: These lists are illustrative, not exhaustive, and identification of noise sources is an active and expanding area of research.

Intrinsic Noise

- *Low copy-number effects (including discrete birth and death events)*: the random nature of production and degradation of cellular components means that noise is high for components at low copy number (as the magnitude of these random fluctuations is not negligible with respect to the copy number);

- *Diffusive cellular dynamics*: many important cellular processes rely on collisions between reactants (for example, RNA polymerase and DNA) and other physical criteria which, given the diffusive dynamic nature of the cell, occur stochastically.

- *Noise propagation*: Low copy-number effects and diffusive dynamics result in each of the biochemical reactions in a cell occurring randomly. Stochasticity of reactions can be either attenuated or amplified. Contribution each reaction makes to the intrinsic variability in copy numbers can be quantified via Van Kampen's system size expansion.

Extrinsic Noise

- *Cellular age / cell cycle stage*: cells in a dividing population that is not synchronised will, at a given snapshot in time, be at different cell cycle stages, with corresponding biochemical and physical differences;

- *Physical environment (temperature, pressure,...)*: physical quantities and chemical concentrations (particularly in the case of cell-to-cell signalling) may vary spatially across a population of cells, provoking extrinsic differences as a function of position;

- *Organelle distributions*: random factors in the quantity and quality of organelles (for example, the number and functionality of mitochondria) lead to significant cell-to-cell differences in a range of processes (as, for example, mitochondria play a central role in the energy budget of eukaryotic cells);

- *Inheritance noise*: uneven partitioning of cellular components between daughter cells at mitosis can result in large extrinsic differences in a dividing population.

Note that extrinsic noise can affect levels and types of intrinsic noise: for example, extrinsic differences in the mitochondrial content of cells lead, through differences in ATP levels, to some cells transcribing faster than others, affecting the rates of gene expression and the magnitude of intrinsic noise across the population.

Effects

Note: These lists are illustrative, not exhaustive, and identification of noise effects is an active and expanding area of research.

- *Gene expression levels*: noise in gene expression causes differences in the fundamental properties of cells, limits their ability to biochemically control cellular dynamics, and directly or indirectly induce many of the specific effects below;

- *Energy levels and transcription rate*: noise in transcription rate, arising from sources including transcriptional bursting, is a significant source of noise in expression levels of genes. Extrinsic noise in mitochondrial content has been suggested to propagate to differences in the ATP concentrations and transcription rates (with functional relationships implied between these three quantities) in cells, affecting cells' energetic competence and ability to express genes;

- *Phenotype selection*: bacterial populations exploit extrinsic noise to choose a population subset to enter a quiescent state. In a bacterial infection, for example, this subset will not propagate quickly but will be more robust when the population is threatened by antibiotic treatment: the rapidly replicating, infectious bacteria will be killed more quickly than the quiescent subset, which may be capable of restarting the infection. This phenomenon is why courses of antibiotics should be finished even when symptoms seem to have disappeared;

- *Development and stem cell differentiation*: developmental noise in biochemical processes which need to be tightly controlled (for example, patterning of gene expression levels that develop into different body parts) during organismal development can have dramatic consequences, necessitating the evolution of robust cellular machinery. Stem cells differentiate into different cell types depending on the expression levels of various characteristic genes: noise in gene expression can clearly perturb and influence this process, and noise in transcription rate can affect the structure of the dynamic landscape that differentiation occurs on;

- *Cancer treatments*: recent work has found extrinsic differences, linked to gene expression levels, in the response of cancer cells to anti-cancer treatments, potentially linking the phenomenon of fractional killing (whereby each treatment kills some but not all of a tumour) to noise in gene expression. Because individual cells could repeatedly and stochastically perform transitions between states associated with differences in responsiveness to a therapeutic modality (chemotherapy, targeted agent, radiation, etc.), therapy might need to be administered frequently (to ensure cells are treated soon after entering a therapy-responsive state, before they can rejoin the therapy-resistant subpopulation and proliferate) and over long times (to treat even those cells emerging late from the final residue of the therapy-resistant subpopulation).

- *Information processing*: as cellular regulation is performed with components that are themselves subject to noise, the ability of cells to process information and perform control is fundamentally limited by intrinsic noise

Analysis

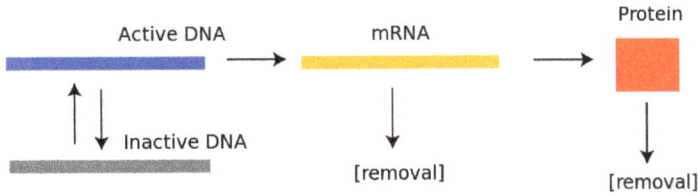

A canonical model for stochastic gene expression. DNA flips between "inactive" and "active" states (involving, for example, chromatin remodelling and transcription factor binding). Active DNA is transcribed to produce mRNA which is translated to produce protein, both of which are degraded. All processes are Poissonian with given rates.

As many quantities of cell biological interest are present in discrete copy number within the cell (single DNAs, dozens of mRNAs, hundreds of proteins), tools from discrete stochastic mathematics are often used to analyse and model cellular noise. In particular, master equation treatments – where the probabilities $P(\mathbf{x},t)$ of observing a system in a state $\mathbf{x}$ at time t are linked through ODEs – have proved particularly fruitful. A canonical model for noise gene expression, where the processes of DNA activation, transcription and translation are all represented as Poisson processes with given rates, gives a master equation which may be solved exactly (with generating functions) under various assumptions or approximated with stochastic tools like Van Kampen's system size expansion.

Numerically, the Gillespie algorithm or stochastic simulation algorithm is often used to create realisations of stochastic cellular processes, from which statistics can be calculated.

The problem of inferring the values of parameters in stochastic models (parametric inference) for biological processes, which are typically characterised by sparse and noisy experimental data, is an active field of research, with methods including Bayesian MCMC and approximate Bayesian computation proving adaptable and robust.

Transcriptional Noise

Transcriptional noise is a primary cause of the variability (noise) in gene expression occurring between cells in isogenic populations. A proposed source of transcriptional noise is transcriptional bursting although other sources of heterogeneity, such as unequal separation of cell contents at mitosis are also likely to contribute considerably. Bursting transcription, as opposed to simple probabilistic models of transcription, reflects multiple states of gene activity, with fluctuations between states separated by irregular intervals, generating uneven protein expression between cells. Noise in gene expression can have tremendous consequences on cell behaviour, and must be mitigated or integrated. In certain contexts, such as the survival of microbes in rapidly changing stressful environments, or several types of scattered differentiation, the variability may be essential. Variability also impacts upon the effectiveness of clinical treatment, with resistance of bacteria to antibiotics demonstrably caused by non-genetic differences. Variability in gene expression may also contribute to resistance of sub-populations of cancer cells to chemotherapy.

Measurement of Expression Noise

The concept of noise in gene expression emerged decades back. But recent advancements in single molecule and single cell analysis methods have allowed the direct observation of noise in organisms. Noise influences the dynamic behavior and functional roles of molecules in a system.

1. It can enable expression of a large set of genes.

2. At the population level, noise provides a wider range of probabilistic differentiation strategy form bacteria to multi cellular organism.

3. Noise facilitates evolutionary adaptation and developmental evolution.

One can characterize the gene expression noise by understanding how protein levels are distributed in individual cells and by the time scale of fluctuations. Three major concepts have evolved to fit our understanding on gene expression noise to an elegant frame work based on experimental and theoretical evidence.

The concept one is related to 'bursts'. The synthesis of proteins during translation is a well quoted event to describe bursts. In a translation event, proteins are not synthesized at a uniform rate, but are produced in stochastic bursts. This might occur because each individual messenger RNA is typically translated several times to synthesize several proteins. It may also be because of stochastic switching of the gene's promoter between long lived OFF and ON state that leads to mRNA bursts which amplify to generate corresponding protein pulse. Such bursts are exhibited by bacteria, yeast, developing embryos and mammalian cells.

Concept two deals with 'time averaging'. When the life time of the protein is longer than the intervals between bursts of protein production, the accumulation of proteins over time averages out the fluctuations generated by the expression of bursts, buffering the protein concentration.

Concept three is related to 'propagation'. The levels of transcription factors and upstream components are subject to bursting and hence influence the rate of gene expression. Hence fluctuations in the expression of a single gene cascades to generate fluctuations in downstream genes. For example, in bacteria slow upstream fluctuations in gene expression rates lead to cellular memory over the time scales of cell cycle regulation.

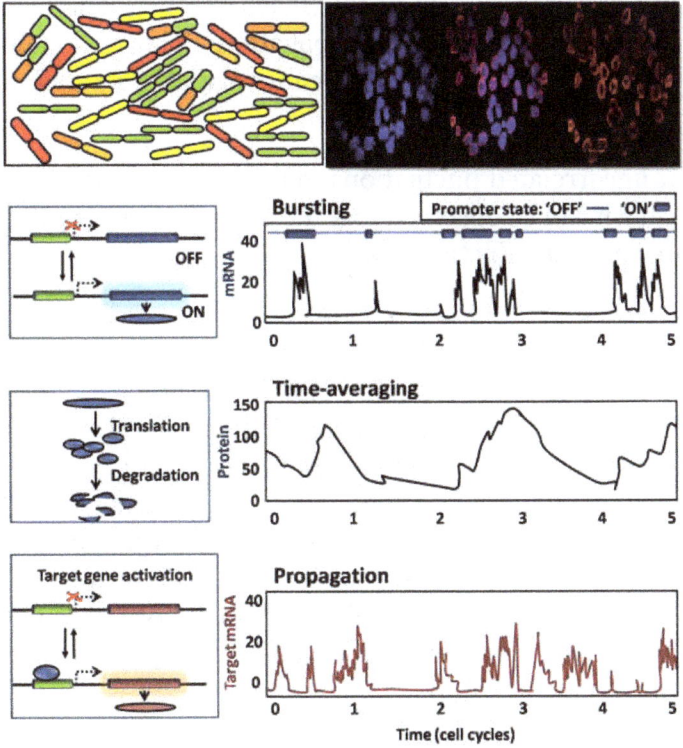

(a) A schematic depicting the various points in the transcription and translation cycle that become sources of noise in gene expression.

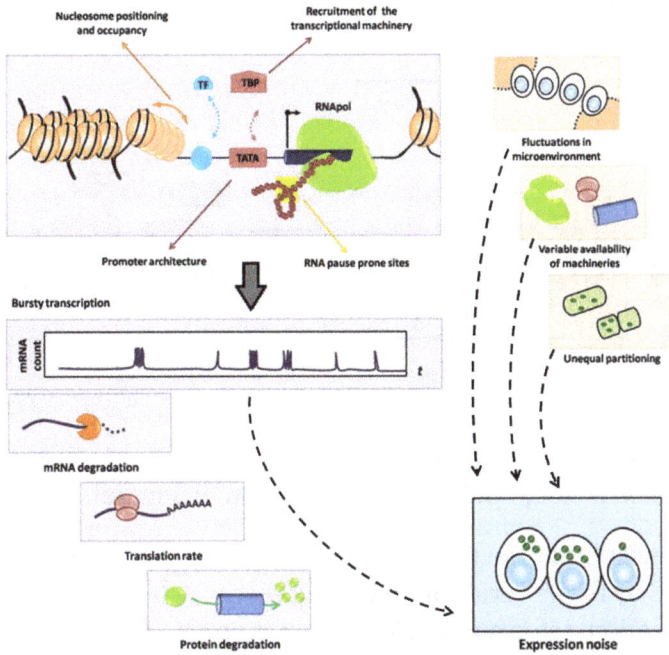

(b). A schematic depicting the various points in the transcription and translation cycle that become sources of noise in gene expression.

It is important to quantify through visualization, the intrinsic noise (noise generating bursts) vs. extrinsic noise (noise propagation). This is done by analyzing the expression of two distinguishable but identically regulated fluorescent reports in the same cell.

One observes that uncorrelated fluctuations in the system occurs due to bursting and time averaging, while it can be observed that the propagation of upstream components are identified with correlated fluctuations.

Quantifying Noise in Gene Expression

The noise component in gene expression arises due to fluctuations in amplitude, frequency, timing of gene expression and stochasticity during the process. As stated already, single molecule techniques offer the sophistication of quantifying the gene expression and to understand the importance of such measurements. At the experimental level, we can attempt to quantify noise by fluorescently tagging the protein of interest. (With GFP (Green Fluorescence Protein) or RFP (Red Fluorescence Protein) or with fluorophores). We can then measure the distribution of normalized fluorescent intensities of this protein across a population of clonal cells in a stable environment. This is clearly explained in Figure.

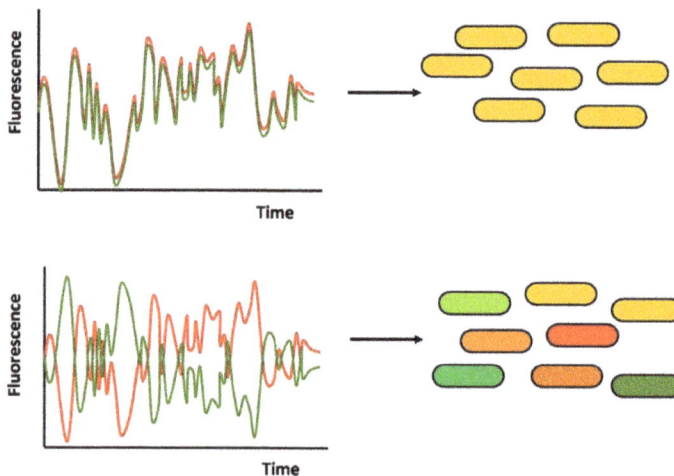

Quantifying gene expression noise using fluorescent reporters

The Figure shows the distribution of normalized fluorescent intensity of the protein of interest across a clonal cell population in a stable environment.

If the standard deviation of the expression levels is denoted by α and the mean expression level denoted by β, the variation in the fluorescence levels is given by $V = \alpha / \beta$. The square of this variation defines noise. Therefore noise $N = (\alpha / \beta)^2$.

The noise due to gene expression can be classified into contributions due to extrinsic and extrinsic components in the gene expression circuits. Hence to measure these two components of a gene x, one uses a dual tag system (double reporter) which is designed in such a way that two identical promoters regulate two distinct fluorescent reporter genes as xGFP and xRFP in this case as shown in the Figure. While quantifying intrinsic noise, one can observe that the fluctuations of each fluorophore are independent of the other. The fluctuations in the fluorescence of GFP and RFP vary dependent on one another while measuring extrinsic noise

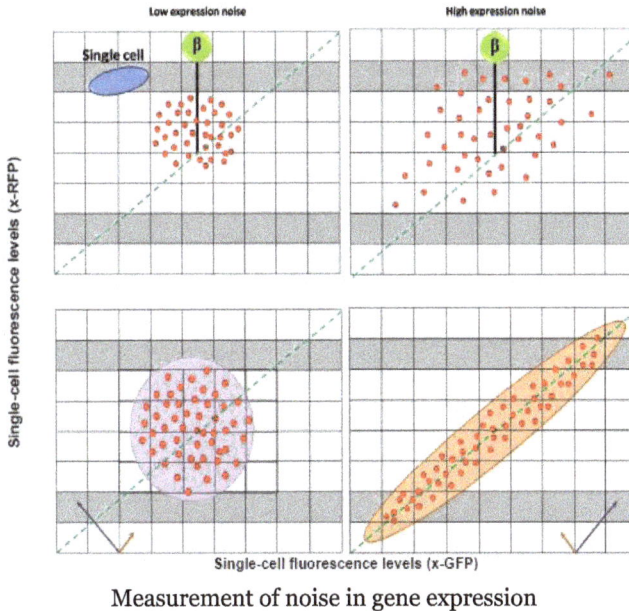

Measurement of noise in gene expression

How do Regulatory Networks Utilise Noise?

Noise is always not an undesirable part of a biological system but defines a core component of how the system develops functions and evolves. Though experiments in the past few years have helped us understand the concept of 'gene expression bursts' in bacteria and eukaryotes, the influence of further layers of regulation on noise remains to be explored. mRNAs have been proven to regulate diverse functions during gene expression, substantially influencing the noise characteristics of its regulated genes (Mehta et al., *Molecular Systems Biology*, 2008, Levine et al., *PLOS Biology*, 2007). This is because of the direct interaction of miRNA with mRNA targets that produce bursts. We thus can explore interesting connections between noise, the architecture of genetic circuits and their biological functions. Hence noise is embedded in bacterial and eukaryotic systems in evolutionary transitions, in developmental systems and in differentiation programs as a key player regulating cellular activities.

Noise in Gene Regulatory Networks

There are inherent stochastic fluctuations during gene expression either at the level of transcription or at the level of translation or during both the events. These fluctuations form the noise components in the systems considered. Despite variations in their levels of expression due to these fluctuations, the genes continue to be functional in the circuit. It is hence important to understand how cells integrate and transmit information through gene regulatory interactions amidst these fluctuations. We now focus on the role of Gene Regulatory Network (GRN) as a buffer and as an amplifier or as a toleration element of gene expression noise. We shall understand the influence of expression noise at distinct levels of network organization- from a simple regulatory network to genome scale regulatory networks.

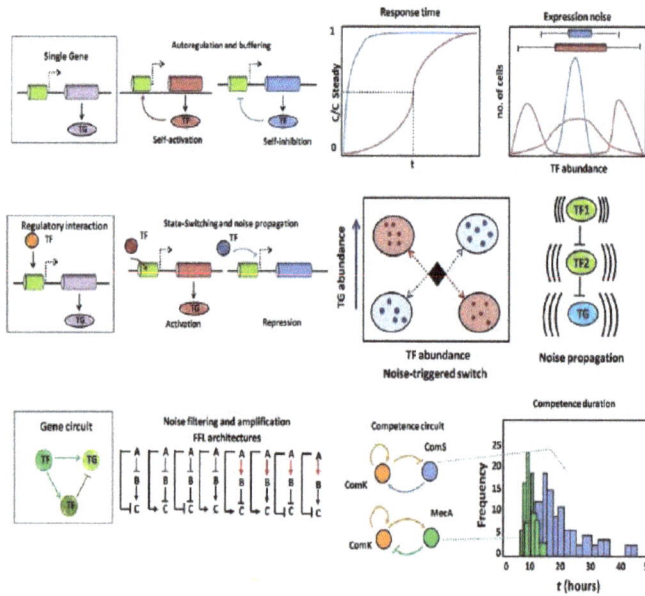

(a)-(c) Expression noise and regulatory networks at distinct levels of network organisation

Noise in Auto Regulation

An auto regulatory network is a simple feedback network, in which a transcription factor binds to its own promoter region regulating its own expression. When the expression pattern is self inhibitory it is called negative auto regulation and when it leads to self activation, it is called positive auto regulation as shown in Figure (b).

Self inhibitory transcription factors are predominantly found in prokaryotic regulatory networks like that of *E.coli*. This negative feedback loop helps the self inhibitory transcription factors in *E.coli* to maintain steady expression levels by controlling the largely the bursts that occur during the self transcription mechanism, which also minimizes the response time (the time required to attain the steady state expression levels). The time taken by a feedback loop to reach the activation threshold shows a sigmoidal response over time and influences gene expression noise. Auto regulatory loops with longer time delays and feedback loops made of more than one component are also known to show enhanced noise.

Noise, Interactions in Regulatory Circuits and State Switching

We now move on from an auto regulation circuit which alters the response time and noise sensitivity of single genes to complex regulatory networks that demand understanding at higher scales of network architecture. A Gene Regulatory Network (GRN) is represented by an interaction between a transcription factor (TF) and a target gene (TG), as in Figure (c).

Fluctuations in transcription factor levels or of the signaling molecules that regulate TF activity result in noise in the target genes, leading to 'noise propagation'. Even slight variations around the critical threshold in transcription factor abundance might lead to a radical switching between distinct states of gene expression, thus yielding distinct cellular phenotypes.

Noise propagation has also been shown to increase with the size of linear regulatory cascades. It assumes a signifi cant role in deciding the information capacity of a regulatory interaction. (Information capacity refers to the number of distinct, stable states in the expression level of a target gene that can be obtained by varying the concentration of a transcription factor). This is better explained in the recent work by Cheong R. et al., (Science 2011) in their study involving the Tumor Necrosis Factor-Nuclear FactorκB (TNF-NFκB) model which suggested that large signalling networks with multiple genes face limitations in the information gained, because of upstream bottle necks in the circuit. Negative feedback to the bottleneck has been shown to decrease noise and also alleviate the limiting effect.

Interplay of Noise and Network Topology- Uniform Stimulus yielding to Differential Cell Fate Outcome

It is an established fact that selective drug treatments during diseased conditions (such as cancer) results in partial survival or significant cell death in clonal cell populations. In such cases the cancer cells behave like bacterial 'persisters' (cells which are capable of driving the survival of a population of cells while treated with antibiotic). Erlotinib is an Epidermal Growth Factor Receptor (EGFR) inhibitor which is used to treat patients with lung adeno carcinoma. It has been found that relapse occurs in patients treated with this drug, due to a minute fraction of non dividing persisters. This tolerance to drug has been shown to originate even in early stages from altered chromatin configurations due to noise in expression of chromatin modifiers. Persisters in lung cancer cells show an increased expression of a type of Histone Demethylase following upstream activation of Insulin like growth factor-1. A combinatorial treatment using IGF-1 receptor inhibitor (to prevent altered chromatin states) followed by Erlotinib treatment recorded an improvement in tumor cell eradication. This clearly emphasizes on understanding the molecular origins of phenotypic variability in isogenic populations while developing treatment strategies.

The expression of a stochastic receptor determines cell fate beyond the action of an interferon treatment on the cells. Presented with interferon, cells show either anti proliferative activity or antiviral response. Since low number of receptors are enough to illicit an anti viral response, it becomes a robust feature of all cells. Cells with higher number of receptors mediate anti proliferative activity. Thus the response for a given cell is binary and is dependent on the stochastic and threshold expression levels of its receptors.

Stem Cell Differentiation is Probabilistic

Stem cells possess properties of both cell renewal (proliferation) and differentiation, and hence regulate cell type and cell number during the development of an organism. The choice of cell fate occurs in a precise and stochastic process. The best example is that of differentiation of neuronal cells where only a fraction of cells in an equivalent population adopts the neuronal cell fate. Shah et al., (*Cell*, 1996) have shown that signals of the Transforming Growth Factor β (TGF β) controls the size of this fraction. High resolution imaging has established that the stem cell state is dynamic and heterogeneous where the cell-cell variability regulates cell fate decision in response to stimuli. In the early mouse embryo, the Inner Cell Mass (ICM) gives rise to distinct lineages like epiblast and primitive endoderm. The epiblast which develops into embryonic tissues expresses Nanog, a regulator of pluripotency. The primitive endoderm produces extra embryonic tissue and expresses Gata6, a transcription factor.

It has been shown that individual cells in the embryo express only one of the transcription factors heterogeneously before the morphological separation between the two fates and signalling through the FGF pathway biases the relative frequency of activity of the two states. This stochasticity clearly shows that the developmental patterning in the mouse embryo is noise dependent.

Here cell fate decisions, though stochastic are not organized spatially at the initial stage. Subsequent spatial arrangement is generated when sorting out of similar type cells takes place. Whether this stochastic process is autonomous or subject to binomial fluctuations in the number of cells of each fate remains to be understood.

Noise in Bacillus Subtilis

The role of noise is also evident in *Bacillus subtilis*. Kuchina et al., (*Molecular Systems Biology*, 2011) have shown a temporal competition between differentiation programs in cell fate decisions in *Bacillus subtilis*. Here the competence and sporulation pathways

are both active and compete with one another till one of the two attains the commitment point. Noise in the expression level of the competing transcription factors regulates the cellular outcome in *Bacillus subtilis* regulatory programs.

Conclusion

The influence of noise on gene regulatory networks on governing inter cellular communications, on defining patterning during development on driving host pathogen interactions can help address interesting questions in biological systems. RNA Fluorescence *insitu* Hybridisation (RNA-FISH), high resolution imaging and single cell transcription, profiling enhanced use of fluorescent proteins may help picturise the dynamic variations in cellular states and help us understand the effect of noise in intracellular regulation and in dynamic switching among states or sub states.

Noise based Switches and Amplifiers for Gene Expression

In the previous discussion, we outlined the influence of noise in gene regulatory networks and cited examples on how external noise controls the dynamics of regulatory networks. We discussed the role of noise in the development of mouse embryo, in *Bacillus subtilis* and in stem cell differentiation. It would be interesting to take an engineering approach to design a genetic regulatory network and manipulate it to control the dynamics of the network. The paper to be discussed now explores an exciting possibility of designing genetic switches and amplifiers to practically control the dynamics of a gene regulatory network. This paper deals with the genetic switch of a lambda phage which makes a decision between lysis and lysogeny and functions as a classic genetic switch.

The work focusses on a network derived from the lambda phage promoter region with a multimeric transcription factor controlling the network. The solitary gene network explored here is a component of a plasmid or a genetic applet which offers distinct advantages. The design confers to an inherent reduction approach reducing the complexity in the gene network making it amenable to a mathematical formulation.

As said earlier, this work discusses an applet derived from the promoter region of the lambda phage. The authors have developed a model with variation in protein concentration in the genetic applet, study the dynamics of the applet under varying protein concentrations. The engineering design is to understand how external noise can be employed to control the network.

Lambda Phage model for Repressor Expression

We know that the bacteriophage – the lambda virus undergoes a lysis or lysogeny pathway and the lambda repressor undergoes auto regulation in the dynamic process. This paper presents two models describing the regulation of the network. The authors propose a system in the form of a plasmid containing P_R- P_{RM} operating region and the components involved in transcription, translation and degradation.

The entire promoter region in lambda phage comprises three operator sites OR1, OR2 and OR3. Initially, the authors describe a mutant system which does not involve the operator site OR1, study the dynamics of the network and categorize the biochemical reactions. They introduce the gene cI to express the repressor CI which dimerizes and binds to the DNA as a transcription factor. Since the mutant system does not involve the operator site OR1 the binding of cI can be only with OR2 or OR3. We take that the cI binding at OR2 increases transcription downstream of OR3 while the cI3 binding at OR3 represses transcription. This repression in transcription turns off production in the circuit.

The authors define two types of chemical reactions- the fast and the slow reaction to describe the network. The fast reactions whose rate constants are in the order of seconds are assumed to be in equilibrium with respect to the slow reactions whose rate constants are in the rate of minutes. Considering the following

X as repressor

X_2 as repressor dimer

D as DNA promoter site

They describe equilibrium reactions

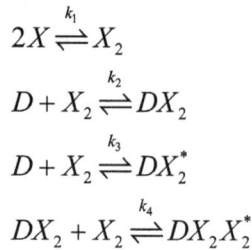

$$2X \overset{k_1}{\rightleftharpoons} X_2$$

$$D + X_2 \overset{k_2}{\rightleftharpoons} DX_2$$

$$D + X_2 \overset{k_3}{\rightleftharpoons} DX_2^*$$

$$DX_2 + X_2 \overset{k_4}{\rightleftharpoons} DX_2 X_2^*$$

Where DX_2 denoted binding to OR2 and DX_2^* denoted binding to OR3. $DX_2 X_2^*$ denotes binding to both sites.

Ki is the forward equilibrium constant. σ_1 and σ_2 represents the binding strengths relative to dimer OR2 strength.

$K_3 = \sigma_1 K_2$

$K_4 = \sigma_2 K_2$

Solving the equations set up to represent the model, the authors observed a bistability in the system.

Considering the effect of the operator site OR1, one finds an increased bistability in the system. They hypothesize that the occurrence of a larger bistable region is to make the switching mechanism more robust to noise and strike a relevance to this observation to the lysis-lysogeny switching mechanism in the lambda phage.

Additive and Multiplicative Noise

This paper then studies the influence of additive external noise on repressor production and its possible relation to bistability. Let the repressor concentration within a cell colony be represented by the dynamic variable X, noise now influences copies of the cell colonies. In the absence of noise, each colony develops identically to the two fixed points described in the model, but this behaviour is modified with an inclusion of noise which brings in fluctuations between colonies. Therefore the noise component which alters the production of the background repressor is called the additive noise.

In the same way, the paper defines multiplicative noise as the one which influences the transcription rate. Since transcription involves a complex sequence of reactions, the introduction of any noise will influence the fluctuations of both internal and external parameters that govern transcription. This type of multiplicative noise can be employed to enhance protein production from a zero concentration level at low noise strength D to a highly populated lower state at low overall concentration. The extension of this concept of noise base switches and amplifiers for gene expression to clinical applications would be interesting in specific applications like gene therapy which requires control systematic regulation of the transfected gene during disease conditions.

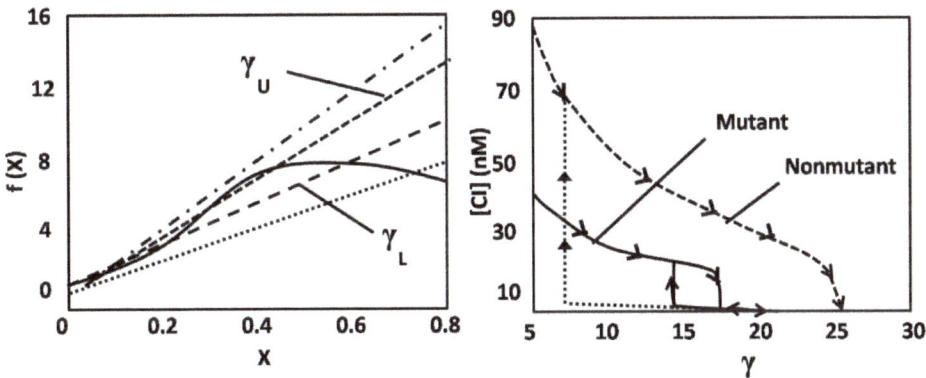

Plot of the variable X with the concentration of the lambda repressor, the mutant and non mutant systems involving the operator site OR1 is shown here.

We can observer that with the increase in slope γ, the system becomes multi stable and returns to a mono stable state.

References

- Fisher, RA; Balmukand, B (July 1928). "The estimation of linkage from the offspring of selfed heterozygotes". Journal of Genetics. 20 (1): 79–92. doi:10.1007/BF02983317

- Griffiths AJF; Miller JH; Suzuki DT; Lewontin RC; et al. (1993). "Chapter 5". An Introduction to Genetic Analysis (5th ed.). New York: W.H. Freeman and Company. ISBN 0-7167-2285-2

- Lobo, Ingrid; Shaw, Kenna. "Discovery and Types of Genetic Linkage". Scitable. Nature Education. Retrieved 21 January 2017

- Morton NE (1955). "Sequential tests for the detection of linkage". American Journal of Human Genetics. 7 (3): 277–318. PMC 1716611. PMID 13258560

- Poehlman JM; Sleper DA (1995). "Chapter 3". Breeding Field Crops (4th ed.). Iowa: Iowa State Press. ISBN 0-8138-2427-3

- Bateson, W; Saunders, ER; Punnett, RC (18 May 1904). Reports to the Evolution committee of the Royal Society. London: Harrison and Sons, Printers. Retrieved 21 January 2017

- Nyholt, Dale R (August 2000). "All LODs Are Not Created Equal". American Journal of Human Genetics. 67 (2): 282–288. doi:10.1086/303029. PMC 1287176. PMID 10884360

- Risch, Neil (June 1991). "A Note on Multiple Testing Procedures in Linkage Analysis" (PDF). American Journal of Human Genetics. 48: 1058–1064. PMC 1683115. PMID 2035526. Retrieved 21 January 2017

- Griffiths, AJF; Miller, JH; Suzuki, DT (2000). "Accurate calculation of large map distances, Derivation of a mapping function". An Introduction to Genetic Analysis (7th ed.). New York: W. H. Freeman. ISBN 0-7167-3520-2

Enzyme Kinetics Models

Enzyme kinetics studies those chemical reactions that are catalyzed by enzymes. Enzymatic reactions speed up the reaction without affecting the reaction equilibrium and hence are important. The aspects elucidated in this chapter are of vital importance, and provide a better understanding of systems biology.

Enzyme Kinetics

Enzyme kinetics is the study of the chemical reactions that are catalysed by enzymes. In enzyme kinetics, the reaction rate is measured and the effects of varying the conditions of the reaction are investigated. Studying an enzyme's kinetics in this way can reveal the catalytic mechanism of this enzyme, its role in metabolism, how its activity is controlled, and how a drug or an agonist might inhibit the enzyme.

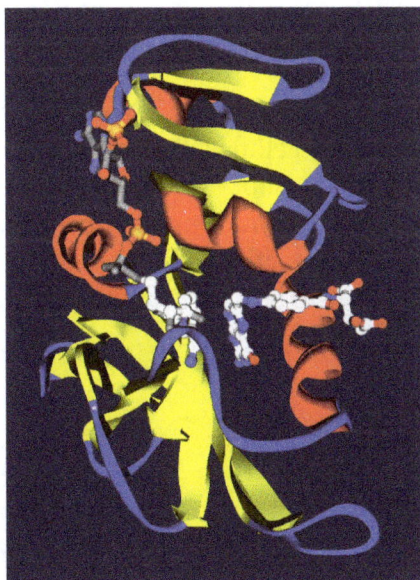

Dihydrofolate reductase from *E. coli* with its two substrates dihydrofolate (right) and NADPH (left), bound in the active site. The protein is shown as a ribbon diagram, with alpha helices in red, beta sheets in yellow and loops in blue. Generated from 7DFR.

Enzymes are usually protein molecules that manipulate other molecules — the enzymes' substrates. These target molecules bind to an enzyme's active site and are transformed into products through a series of steps known as the enzymatic mechanism

$$E + S \rightleftarrows ES \rightleftarrows ES^* \rightleftarrows EP \rightleftarrows E + P$$

These mechanisms can be divided into single-substrate and multiple-substrate mechanisms. Kinetic studies on enzymes that only bind one substrate, such as triosephosphate isomerase, aim to measure the affinity with which the enzyme binds this substrate and the turnover rate. Some other examples of enzymes are phosphofructokinase and hexokinase, both of which are important for cellular respiration (glycolysis).

When enzymes bind multiple substrates, such as dihydrofolate reductase, enzyme kinetics can also show the sequence in which these substrates bind and the sequence in which products are released. An example of enzymes that bind a single substrate and release multiple products are proteases, which cleave one protein substrate into two polypeptide products. Others join two substrates together, such as DNA polymerase linking a nucleotide to DNA. Although these mechanisms are often a complex series of steps, there is typically one *rate-determining step* that determines the overall kinetics. This rate-determining step may be a chemical reaction or a conformational change of the enzyme or substrates, such as those involved in the release of product(s) from the enzyme.

Knowledge of the enzyme's structure is helpful in interpreting kinetic data. For example, the structure can suggest how substrates and products bind during catalysis; what changes occur during the reaction; and even the role of particular amino acid residues in the mechanism. Some enzymes change shape significantly during the mechanism; in such cases, it is helpful to determine the enzyme structure with and without bound substrate analogues that do not undergo the enzymatic reaction.

Not all biological catalysts are protein enzymes; RNA-based catalysts such as ribozymes and ribosomes are essential to many cellular functions, such as RNA splicing and translation. The main difference between ribozymes and enzymes is that RNA catalysts are composed of nucleotides, whereas enzymes are composed of amino acids. Ribozymes also perform a more limited set of reactions, although their reaction mechanisms and kinetics can be analysed and classified by the same methods.

General Principles

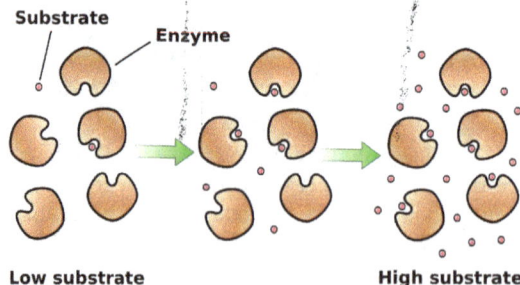

As larger amounts of substrate are added to a reaction, the available enzyme binding sites become filled to the limit of V_{max}. Beyond this limit the enzyme is saturated with substrate and the reaction rate ceases to increase.

The reaction catalysed by an enzyme uses exactly the same reactants and produces exactly the same products as the uncatalysed reaction. Like other catalysts, enzymes do not alter the position of equilibrium between substrates and products. However, unlike uncatalysed chemical reactions, enzyme-catalysed reactions display saturation kinetics. For a given enzyme concentration and for relatively low substrate concentrations, the reaction rate increases linearly with substrate concentration; the enzyme molecules are largely free to catalyse the reaction, and increasing substrate concentration means an increasing rate at which the enzyme and substrate molecules encounter one another. However, at relatively high substrate concentrations, the reaction rate asymptotically approaches the theoretical maximum; the enzyme active sites are almost all occupied and the reaction rate is determined by the intrinsic turnover rate of the enzyme. The substrate concentration midway between these two limiting cases is denoted by K_M.

The two most important kinetic properties of an enzyme are how quickly the enzyme becomes saturated with a particular substrate, and the maximum rate it can achieve. Knowing these properties suggests what an enzyme might do in the cell and can show how the enzyme will respond to changes in these conditions.

Enzyme Assays

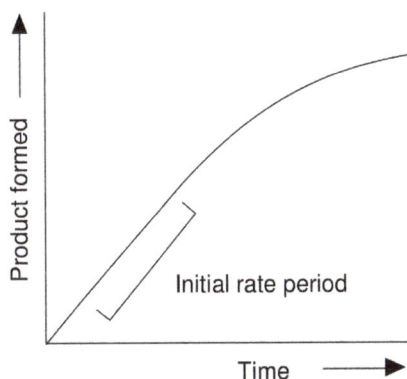

Progress curve for an enzyme reaction. The slope in the initial rate period is the initial rate of reaction v. The Michaelis–Menten equation describes how this slope varies with the concentration of substrate.

Enzyme assays are laboratory procedures that measure the rate of enzyme reactions. Because enzymes are not consumed by the reactions they catalyse, enzyme assays usually follow changes in the concentration of either substrates or products to measure the rate of reaction. There are many methods of measurement. Spectrophotometric assays observe change in the absorbance of light between products and reactants; radiometric assays involve the incorporation or release of radioactivity to measure the amount of product made over time. Spectrophotometric assays are most convenient since they allow the rate of the reaction to be measured continuously. Although radiometric assays require the removal and counting of samples (i.e., they are discontinuous assays) they are usually extremely sensitive and can measure very low levels of enzyme activity. An analogous approach is to use mass spectrometry to monitor the incorporation or release of stable isotopes as substrate is converted into product.

The most sensitive enzyme assays use lasers focused through a microscope to observe changes in single enzyme molecules as they catalyse their reactions. These measurements either use changes in the fluorescence of cofactors during an enzyme's reaction mechanism, or of fluorescent dyes added onto specific sites of the protein to report movements that occur during catalysis. These studies are providing a new view of the kinetics and dynamics of single enzymes, as opposed to traditional enzyme kinetics, which observes the average behaviour of populations of millions of enzyme molecules.

An example progress curve for an enzyme assay is shown above. The enzyme produces product at an initial rate that is approximately linear for a short period after the start of the reaction. As the reaction proceeds and substrate is consumed, the rate continuously slows (so long as substrate is not still at saturating levels). To measure the initial (and maximal) rate, enzyme assays are typically carried out while the reaction has progressed only a few percent towards total completion. The length of the initial rate period depends on the assay conditions and can range from milliseconds to hours. However, equipment for rapidly mixing liquids allows fast kinetic measurements on initial rates of less than one second. These very rapid assays are essential for measuring pre-steady-state kinetics, which are discussed below.

Most enzyme kinetics studies concentrate on this initial, approximately linear part of enzyme reactions. However, it is also possible to measure the complete reaction curve and fit this data to a non-linear rate equation. This way of measuring enzyme reactions is called progress-curve analysis. This approach is useful as an alternative to rapid kinetics when the initial rate is too fast to measure accurately.

Single-substrate Reactions

Enzymes with single-substrate mechanisms include isomerases such as triosephosphateisomerase or bisphosphoglycerate mutase, intramolecular lyases such as adenylate cyclase and the hammerhead ribozyme, an RNA lyase. However, some enzymes that only have a single substrate do not fall into this category of mechanisms. Catalase is an example of this, as the enzyme reacts with a first molecule of hydrogen peroxide substrate, becomes oxidised and is then reduced by a second molecule of substrate. Although a single substrate is involved, the existence of a modified enzyme intermediate means that the mechanism of catalase is actually a ping–pong mechanism, a type of mechanism that is discussed in the *Multi-substrate reactions* section below.

Michaelis–Menten Kinetics

As enzyme-catalysed reactions are saturable, their rate of catalysis does not show a linear response to increasing substrate. If the initial rate of the reaction is measured over a range of substrate concentrations (denoted as [S]), the reaction rate (v) increases as [S] increases. However, as [S] gets higher, the enzyme becomes saturated with substrate and the rate reaches V_{max}, the enzyme's maximum rate.

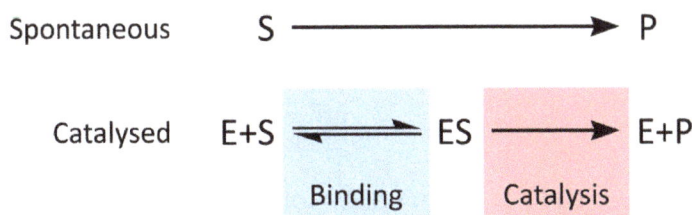

Spontaneous   S ⟶ P

Catalysed   E+S ⇌ ES ⟶ E+P

Binding   Catalysis

A chemical reaction mechanism with or without enzyme catalysis. The enzyme (E) binds substrate (S) to produce product (P).

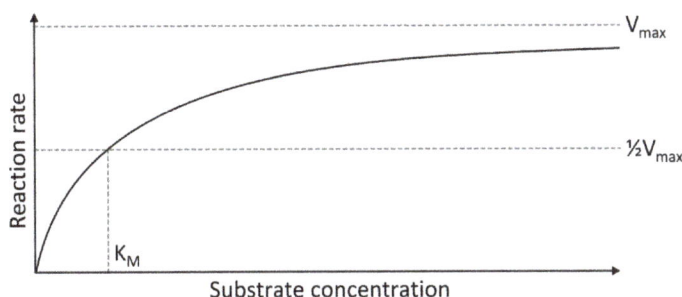

Saturation curve for an enzyme reaction showing the relation between the substrate concentration and reaction rate.

The Michaelis–Menten kinetic model of a single-substrate reaction. There is an initial bimolecular reaction between the enzyme E and substrate S to form the enzyme–substrate complex ES. The rate of enzymatic reaction increases with the increase of the substrate concentration up to a certain level called V_{max}; at V_{max}, increase in substrate concentration does not cause any increase in reaction rate as there no more enzyme (E) available for reacting with substrate (S). Here, the rate of reaction becomes dependent on the ES complex and the reaction becomes a unimolecular reaction with an order of zero. Though the enzymatic mechanism for the unimolecular reaction $ES \xrightarrow{k_{cat}} E+P$ can be quite complex, there is typically one rate-determining enzymatic step that allows this reaction to be modelled as a single catalytic step with an apparent unimolecular rate constant k_{cat}. If the reaction path proceeds over one or several intermediates, k_{cat} will be a function of several elementary rate constants, whereas in the simplest case of a single elementary reaction (e.g. no intermediates) it will be identical to the elementary unimolecular rate constant k_2. The apparent unimolecular rate constant k_{cat} is also called *turnover number* and denotes the maximum number of enzymatic reactions catalysed per second.

The Michaelis–Menten equation describes how the (initial) reaction rate v_0 depends on the position of the substrate-binding equilibrium and the rate constant k_2.

$$v_0 = \frac{V_{max}[S]}{K_M + [S]} \quad \textit{(Michaelis–Menten equation)}$$

with the constants

$$K_M \overset{\text{def}}{=} \frac{k_2 + k_{-1}}{k_1} \approx K_D$$

$$V_{max} \overset{\text{def}}{=} k_{cat}[E]_{tot}$$

This Michaelis–Menten equation is the basis for most single-substrate enzyme kinetics. Two crucial assumptions underlie this equation (apart from the general assumption about the mechanism only involving no intermediate or product inhibition, and there is no allostericity or cooperativity). The first assumption is the so-called quasi-steady-state assumption (or pseudo-steady-state hypothesis), namely that the concentration of the substrate-bound enzyme (and hence also the unbound enzyme) changes much more slowly than those of the product and substrate and thus the change over time of the complex can be set to zero $d[ES]/dt \overset{!}{=} 0$. The second assumption is that the total enzyme concentration does not change over time, thus $[E]_{tot} = [E] + [ES] \overset{!}{=} \text{const}$. A complete derivation can be found here.

The Michaelis constant K_M is experimentally defined as the concentration at which the rate of the enzyme reaction is half V_{max}, which can be verified by substituting $[S] = K_m$ into the Michaelis–Menten equation and can also be seen graphically. If the rate-determining enzymatic step is slow compared to substrate dissociation ($k_2 \ll k_{-1}$), the Michaelis constant K_M is roughly the dissociation constant K_D of the ES complex.

If $[S]$ is small compared to K_M then the term $[S]/(K_M + [S]) \approx [S]/K_M$ and also very little ES complex is formed, thus $[E]_0 \approx [E]$. Therefore, the rate of product formation is

$$v_0 \approx \frac{k_{cat}}{K_M}[E][S] \qquad \text{if } [S] \ll K_M$$

Thus the product formation rate depends on the enzyme concentration as well as on the substrate concentration, the equation resembles a bimolecular reaction with a corresponding pseudo-second order rate constant k_2/K_M. This constant is a measure of catalytic efficiency. The most efficient enzymes reach a k_2/K_M in the range of $10^8 - 10^{10}$ M^{-1} s^{-1}. These enzymes are so efficient they effectively catalyse a reaction each time they encounter a substrate molecule and have thus reached an upper theoretical limit for efficiency (diffusion limit); and are sometimes referred to as kinetically perfect enzymes.

Direct use of the Michaelis–Menten Equation for Time Course Kinetic Analysis

The observed velocities predicted by the Michaelis–Menten equation can be used to directly model the time course disappearance of substrate and the production of product through incorporation of the Michaelis–Menten equation into the equation for first order chemical kinetics. This can only be achieved however if one recognises the problem associated with the use of Euler's number in the description of first order chemical kinetics. i.e. e^{-k} is a split constant that introduces a systematic error into calculations and can be rewritten as a single constant which represents the remaining substrate after each time period.

$$[S] = [S]_0 (1-k)^t$$

$$[S] = [S]_0 (1 - v/[S]_0)^t$$

$$[S] = [S]_0 (1 - (V_{max}[S]_0/(K_M + [S]_0)/[S]_0))^t$$

In 1983 Stuart Beal (and also independently Santiago Schnell and Claudio Mendoza in 1997) derived a closed form solution for the time course kinetics analysis of the Michaelis-Menten mechanism. The solution, known as the Schnell-Mendoza equation, has the form:

$$\frac{[S]}{K_M} = W[F(t)]$$

where W[] is the Lambert-W function. and where F(t) is

$$F(t) = \frac{[S]_0}{K_M} \exp\left(\frac{[S]_0}{K_M} - \frac{V_{max}}{K_M} t \right)$$

This equation is encompassed by the equation below, obtained by Berberan-Santos (MATCH Commun. Math. Comput. Chem. 63 (2010) 283), which is also valid when the initial substrate concentration is close to that of enzyme,

$$\frac{[S]}{K_M} = W[F(t)] - \frac{V_{max}}{k_{cat}K_M} \frac{W[F(t)]}{1 + W[F(t)]}$$

where W[] is again the Lambert-W function.

Linear Plots of the Michaelis–Menten Equation

The plot of v versus [S] above is not linear; although initially linear at low [S], it bends over to saturate at high [S]. Before the modern era of nonlinear curve-fitting on computers, this nonlinearity could make it difficult to estimate K_M and V_{max} accurately. Therefore, several researchers developed linearisations of the Michaelis–Menten equation, such as the Lineweaver–Burk plot, the Eadie–Hofstee diagram and the Hanes–Woolf plot. All of these linear representations can be useful for visualising data, but none should be used to determine kinetic parameters, as computer software is readily available that allows for more accurate determination by nonlinear regression methods.

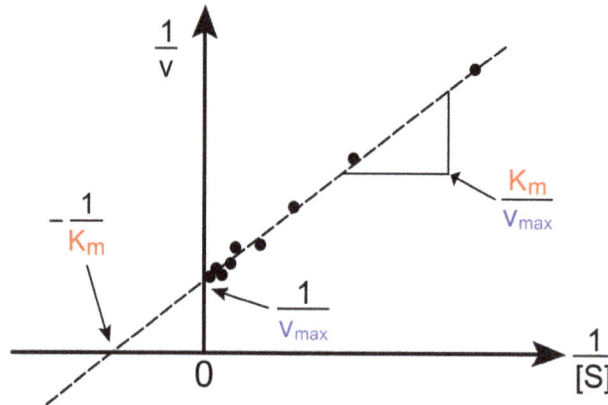

Lineweaver–Burk or double-reciprocal plot of kinetic data, showing
the significance of the axis intercepts and gradient.

The Lineweaver–Burk plot or double reciprocal plot is a common way of illustrating kinetic data. This is produced by taking the reciprocal of both sides of the Michaelis–Menten equation. As shown on the figure, this is a linear form of the Michaelis–Menten equation and produces a straight line with the equation $y = mx + c$ with a y-intercept equivalent to $1/V_{max}$ and an x-intercept of the graph representing $-1/K_M$.

$$\frac{1}{v} = \frac{K_M}{V_{max}[S]} + \frac{1}{V_{max}}$$

Naturally, no experimental values can be taken at negative $1/[S]$; the lower limiting value $1/[S] = 0$ (the y-intercept) corresponds to an infinite substrate concentration, where $1/v = 1/V_{max}$ thus, the x-intercept is an extrapolation of the experimental data taken at positive concentrations. More generally, the Lineweaver–Burk plot skews the importance of measurements taken at low substrate concentrations and, thus, can yield inaccurate estimates of V_{max} and K_M. A more accurate linear plotting method is the Eadie-Hofstee plot. In this case, v is plotted against $v/[S]$. In the third common linear representation, the Hanes-Woolf plot, $[S]/v$ is plotted against $[S]$. In general, data normalisation can help diminish the amount of experimental work and can increase the reliability of the output, and is suitable for both graphical and numerical analysis.

Practical Significance of Kinetic Constants

The study of enzyme kinetics is important for two basic reasons. Firstly, it helps explain how enzymes work, and secondly, it helps predict how enzymes behave in living organisms. The kinetic constants defined above, K_M and V_{max}, are critical to attempts to understand how enzymes work together to control metabolism.

Making these predictions is not trivial, even for simple systems. For example, oxaloacetate is formed by malate dehydrogenase within the mitochondrion. Oxaloacetate can then be consumed by citrate synthase, phosphoenolpyruvate carboxykinase or aspartate aminotransferase, feeding into the citric acid cycle, gluconeogenesis or aspartic acid biosynthesis, respectively. Being able to predict how much oxaloacetate goes into which pathway requires knowledge of the concentration of oxaloacetate as well as the concentration and kinetics of each of these enzymes. This aim of predicting the behaviour of metabolic pathways reaches its most complex expression in the synthesis of huge amounts of kinetic and gene expression data into mathematical models of entire organisms. Alternatively, one useful simplification of the metabolic modelling problem is to ignore the underlying enzyme kinetics and only rely on information about the reaction network's stoichiometry, a technique called flux balance analysis.

Michaelis–Menten Kinetics with Intermediate

One could also consider the less simple case

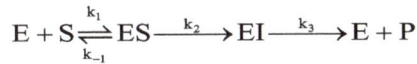

$$E + S \underset{k_{-1}}{\overset{k_1}{\rightleftharpoons}} ES \xrightarrow{k_2} EI \xrightarrow{k_3} E + P$$

where a complex with the enzyme and an intermediate exists and the intermediate is converted into product in a second step. In this case we have a very similar equation

$$v_0 = k_{cat} \frac{[S][E]_0}{K_M' + [S]}$$

but the constants are different

$$K_M' \overset{def}{=} \frac{k_3}{k_2 + k_3} K_M = \frac{k_3}{k_2 + k_3} \cdot \frac{k_2 + k_{-1}}{k_1}$$

$$k_{cat} \overset{def}{=} \frac{k_3 k_2}{k_2 + k_3}$$

We see that for the limiting case $k_3 \gg k_2$, thus when the last step from $EI \to E + P$ is much faster than the previous step, we get again the original equation. Mathematically we have then $K_M' \approx K_M$ and $k_{cat} \approx k_2$.

Multi-substrate Reactions

Multi-substrate reactions follow complex rate equations that describe how the substrates bind and in what sequence. The analysis of these reactions is much simpler if the concentration of substrate A is kept constant and substrate B varied. Under these conditions, the enzyme behaves just like a single-substrate enzyme and a plot of v by [S] gives apparent K_M and V_{max} constants for substrate B. If a set of these measurements is performed at different fixed concentrations of A, these data can be used to work out what the mechanism of the reaction is. For an enzyme that takes two substrates A and B and turns them into two products P and Q, there are two types of mechanism: ternary complex and ping–pong.

Ternary-complex Mechanisms

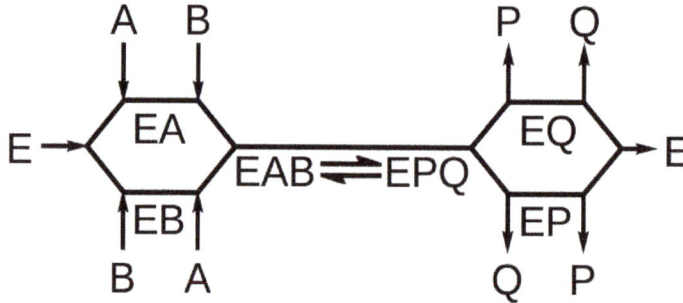

Random-order ternary-complex mechanism for an enzyme reaction. The reaction path is shown as a line and enzyme intermediates containing substrates A and B or products P and Q are written below the line.

In these enzymes, both substrates bind to the enzyme at the same time to produce an EAB ternary complex. The order of binding can either be random (in a random mechanism) or substrates have to bind in a particular sequence (in an ordered mechanism). When a set of v by [S] curves (fixed A, varying B) from an enzyme with a ternary-complex mechanism are plotted in a Lineweaver–Burk plot, the set of lines produced will intersect.

Enzymes with ternary-complex mechanisms include glutathione S-transferase, dihydrofolate reductase and DNA polymerase.

Ping–pong mechanisms

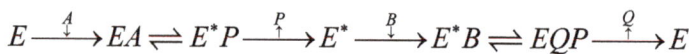

$$E \xrightarrow{A} EA \rightleftharpoons E^*P \xrightarrow{P} E^* \xrightarrow{B} E^*B \rightleftharpoons EQP \xrightarrow{Q} E$$

Ping–pong mechanism for an enzyme reaction. Intermediates contain substrates A and B or products P and Q.

As shown on the above, enzymes with a ping-pong mechanism can exist in two states, E and a chemically modified form of the enzyme E*; this modified enzyme is known as an intermediate. In such mechanisms, substrate A binds, changes the enzyme to E* by, for example, transferring a chemical group to the active site, and is then released. Only after the first substrate is released can substrate B bind and react with the modified enzyme, regenerating the unmodified E form. When a set of v by [S] curves (fixed A, varying B) from an enzyme with a ping–pong mechanism are plotted in a Lineweaver–Burk plot, a set of parallel lines will be produced. This is called a secondary plot.

Enzymes with ping–pong mechanisms include some oxidoreductases such as thiore-doxin peroxidase, transferases such as acylneuraminate cytidylyltransferase and serine proteases such as trypsin and chymotrypsin. Serine proteases are a very common and diverse family of enzymes, including digestive enzymes (trypsin, chymotrypsin, and elastase), several enzymes of the blood clotting cascade and many others. In these ser-ine proteases, the E* intermediate is an acyl-enzyme species formed by the attack of an active site serine residue on a peptide bond in a protein substrate.

Non-Michaelis–Menten Kinetics

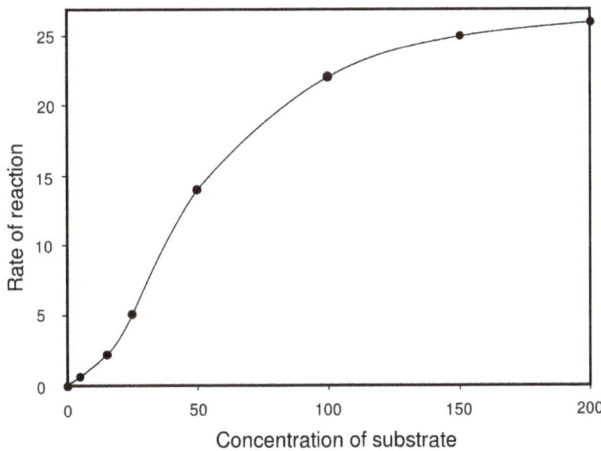

Saturation curve for an enzyme reaction showing sigmoid kinetics.

Some enzymes produce a sigmoid v by [S] plot, which often indicates cooperative binding of substrate to the active site. This means that the binding of one substrate molecule affects the binding of subsequent substrate molecules. This behavior is most common in multimeric enzymes with several interacting active sites. Here, the mechanism of cooperation is similar to that of hemoglobin, with binding of sub-strate to one active site altering the affinity of the other active sites for substrate molecules. Positive cooperativity occurs when binding of the first substrate molecule *increases* the affinity of the other active sites for substrate. Negative cooperativity occurs when binding of the first substrate *decreases* the affinity of the enzyme for other substrate molecules.

Allosteric enzymes include mammalian tyrosyl tRNA-synthetase, which shows neg-ative cooperativity, and bacterial aspartate transcarbamoylase and phosphofructoki-nase, which show positive cooperativity.

Cooperativity is surprisingly common and can help regulate the responses of enzymes to changes in the concentrations of their substrates. Positive cooperativity makes enzymes much more sensitive to [S] and their activities can show large changes over a narrow range of substrate concentration. Conversely, negative cooperativity makes enzymes insensitive to small changes in [S].

The Hill equation (biochemistry) is often used to describe the degree of cooperativity quantitatively in non-Michaelis–Menten kinetics. The derived Hill coefficient n measures how much the binding of substrate to one active site affects the binding of substrate to the other active sites. A Hill coefficient of <1 indicates negative cooperativity and a coefficient of >1 indicates positive cooperativity.

Pre-Steady-State Kinetics

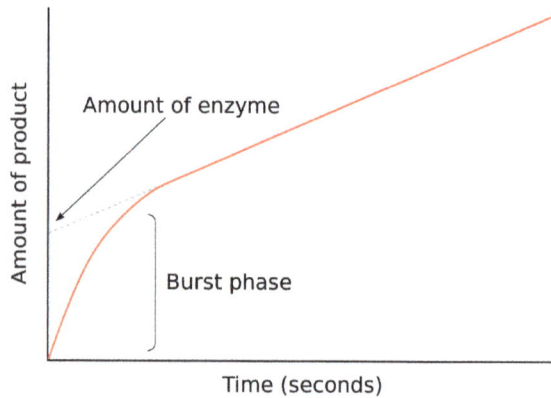

Pre-steady state progress curve, showing the burst phase of an enzyme reaction.

In the first moment after an enzyme is mixed with substrate, no product has been formed and no intermediates exist. The study of the next few milliseconds of the reaction is called pre-steady-state kinetics. Pre-steady-state kinetics is therefore concerned with the formation and consumption of enzyme–substrate intermediates (such as ES or E*) until their steady-state concentrations are reached.

This approach was first applied to the hydrolysis reaction catalysed by chymotrypsin. Often, the detection of an intermediate is a vital piece of evidence in investigations of what mechanism an enzyme follows. For example, in the ping–pong mechanisms that are shown above, rapid kinetic measurements can follow the release of product P and measure the formation of the modified enzyme intermediate E*. In the case of chymotrypsin, this intermediate is formed by an attack on the substrate by the nucleophilic serine in the active site and the formation of the acyl-enzyme intermediate.

In the figure to the right, the enzyme produces E* rapidly in the first few seconds of the reaction. The rate then slows as steady state is reached. This rapid burst phase of the reaction measures a single turnover of the enzyme. Consequently, the amount of product released in this burst, shown as the intercept on the y-axis of the graph, also gives the amount of functional enzyme which is present in the assay.

Chemical Mechanism

An important goal of measuring enzyme kinetics is to determine the chemical mechanism of an enzyme reaction, i.e., the sequence of chemical steps that transform substrate into product. The kinetic approaches discussed above will show at what rates intermediates are formed and inter-converted, but they cannot identify exactly what these intermediates are.

Kinetic measurements taken under various solution conditions or on slightly modified enzymes or substrates often shed light on this chemical mechanism, as they reveal the rate-determining step or intermediates in the reaction. For example, the breaking of a covalent bond to a hydrogen atom is a common rate-determining step. Which of the possible hydrogen transfers is rate determining can be shown by measuring the kinetic effects of substituting each hydrogen by deuterium, its stable isotope. The rate will change when the critical hydrogen is replaced, due to a primary kinetic isotope effect, which occurs because bonds to deuterium are harder to break than bonds to hydrogen. It is also possible to measure similar effects with other isotope substitutions, such as $^{13}C/^{12}C$ and $^{18}O/^{16}O$, but these effects are more subtle.

Isotopes can also be used to reveal the fate of various parts of the substrate molecules in the final products. For example, it is sometimes difficult to discern the origin of an oxygen atom in the final product; since it may have come from water or from part of the substrate. This may be determined by systematically substituting oxygen's stable isotope ^{18}O into the various molecules that participate in the reaction and checking for the isotope in the product. The chemical mechanism can also be elucidated by examining the kinetics and isotope effects under different pH conditions, by altering the metal ions or other bound cofactors, by site-directed mutagenesis of conserved amino acid residues, or by studying the behaviour of the enzyme in the presence of analogues of the substrate(s).

Enzyme Inhibition and Activation

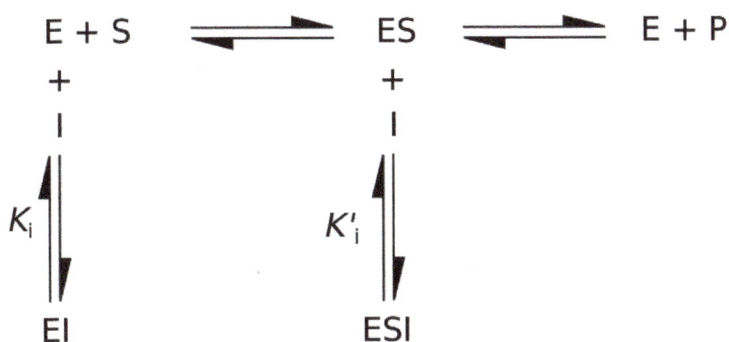

Kinetic scheme for reversible enzyme inhibitors.

Enzyme inhibitors are molecules that reduce or abolish enzyme activity, while enzyme activators are molecules that increase the catalytic rate of enzymes. These interactions can be either *reversible* (i.e., removal of the inhibitor restores enzyme activity) or *irreversible* (i.e., the inhibitor permanently inactivates the enzyme).

Reversible Inhibitors

Traditionally reversible enzyme inhibitors have been classified as competitive, uncompetitive, or non-competitive, according to their effects on K_m and V_{max}. These different effects result from the inhibitor binding to the enzyme E, to the enzyme–substrate complex ES, or to both, respectively. The division of these classes arises from a problem in their derivation and results in the need to use two different binding constants for one binding event. The binding of an inhibitor and its effect on the enzymatic activity are two distinctly different things, another problem the traditional equations fail to acknowledge. In noncompetitive inhibition the binding of the inhibitor results in 100% inhibition of the enzyme only, and fails to consider the possibility of anything in between. The common form of the inhibitory term also obscures the relationship between the inhibitor binding to the enzyme and its relationship to any other binding term be it the Michaelis–Menten equation or a dose response curve associated with ligand receptor binding. To demonstrate the relationship the following rearrangement can be made:

$$\frac{V_{max}}{1+\frac{[I]}{K_i}} = \frac{V_{max}}{\frac{[I]+K_i}{K_i}}$$

Adding zero to the bottom ([I]-[I])

$$\frac{\frac{V_{max}}{[I]+K_i}}{[I]+K_i-[I]}$$

Dividing by [I]+K$_i$

$$\frac{\frac{V_{max}}{1}}{1-\frac{[I]}{[I]+K_i}} = V_{max} - V_{max}\frac{[I]}{[I]+K_i}$$

This notation demonstrates that similar to the Michaelis–Menten equation, where the rate of reaction depends on the percent of the enzyme population interacting with substrate

fraction of the enzyme population bound by substrate

$$\frac{[S]}{[S] + K_m}$$

fraction of the enzyme population bound by inhibitor

$$\frac{[I]}{[I] + K_i}$$

the effect of the inhibitor is a result of the percent of the enzyme population interacting with inhibitor. The only problem with this equation in its present form is that it assumes absolute inhibition of the enzyme with inhibitor binding, when in fact there can be a wide range of effects anywhere from 100% inhibition of substrate turn over to just >0%. To account for this the equation can be easily modified to allow for different degrees of inhibition by including a delta V_{max} term.

$$V_{max} - \Delta V_{max} \frac{[I]}{[I] + K_i}$$

or

$$V_{max1} - (V_{max1} - V_{max2}) \frac{[I]}{[I] + K_i}$$

This term can then define the residual enzymatic activity present when the inhibitor is interacting with individual enzymes in the population. However the inclusion of this term has the added value of allowing for the possibility of activation if the secondary V_{max} term turns out to be higher than the initial term. To account for the possibly of activation as well the notation can then be rewritten replacing the inhibitor "I" with a modifier term denoted here as "X".

$$V_{max1} - (V_{max1} - V_{max2}) \frac{[X]}{[X] + K_x}$$

While this terminology results in a simplified way of dealing with kinetic effects relating to the maximum velocity of the Michaelis–Menten equation, it highlights potential problems with the term used to describe effects relating to the K_m. The K_m relating to the affinity of the enzyme for the substrate should in most cases relate to potential changes in the binding site of the enzyme which would directly result from enzyme inhibitor interactions. As such a term similar to the one proposed above to modulate V_{max} should be appropriate in most situations:

$$K_{m1} - (K_{m1} - K_{m2}) \frac{[X]}{[X] + K_x}$$

Irreversible Inhibitors

Enzyme inhibitors can also irreversibly inactivate enzymes, usually by covalently modifying active site residues. These reactions, which may be called suicide substrates, follow exponential decay functions and are usually saturable. Below saturation, they follow first order kinetics with respect to inhibitor.

Mechanisms of Catalysis

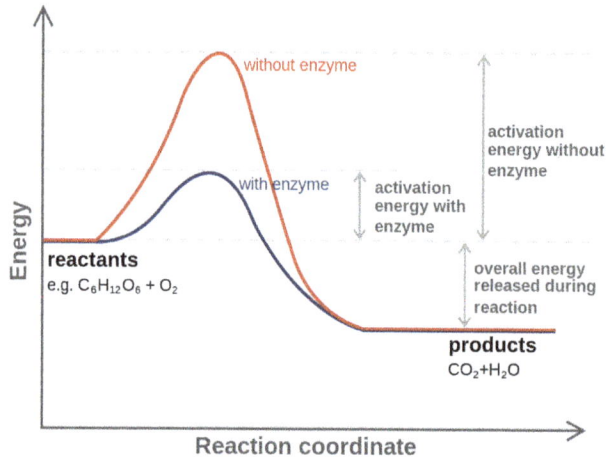

The energy variation as a function of reaction coordinate shows the stabilisation of the transition state by an enzyme.

The favoured model for the enzyme–substrate interaction is the induced fit model. This model proposes that the initial interaction between enzyme and substrate is relatively weak, but that these weak interactions rapidly induce conformational changes in the enzyme that strengthen binding. These conformational changes also bring catalytic residues in the active site close to the chemical bonds in the substrate that will be altered in the reaction. Conformational changes can be measured using circular dichroism or dual polarisation interferometry. After binding takes place, one or more mechanisms of catalysis lower the energy of the reaction's transition state by providing an alternative chemical pathway for the reaction. Mechanisms of catalysis include catalysis by bond strain; by proximity and orientation; by active-site proton donors or acceptors; covalent catalysis and quantum tunnelling.

Enzyme kinetics cannot prove which modes of catalysis are used by an enzyme. However, some kinetic data can suggest possibilities to be examined by other techniques. For example, a ping–pong mechanism with burst-phase pre-steady-state kinetics would suggest covalent catalysis might be important in this enzyme's mechanism. Alternatively, the observation of a strong pH effect on V_{max} but not K_m might indicate that a residue in the active site needs to be in a particular ionisation state for catalysis to occur.

History

In 1902 Victor Henri proposed a quantitative theory of enzyme kinetics, but at the time the experimental significance of the hydrogen ion concentration was not yet recognized. After Peter Lauritz Sørensen had defined the logarithmic pH-scale and introduced the concept of buffering in 1909 the German chemist Leonor Michaelis and his Canadian postdoc Maud Leonora Menten repeated Henri's experiments and confirmed his equation, which is now generally referred to as Michaelis-Menten kinetics (sometimes also *Henri-Michaelis-Menten kinetics*). Their work was further developed by G. E. Briggs and J. B. S. Haldane, who derived kinetic equations that are still widely considered today a starting point in modeling enzymatic activity.

The major contribution of the Henri-Michaelis-Menten approach was to think of enzyme reactions in two stages. In the first, the substrate binds reversibly to the enzyme, forming the enzyme-substrate complex. This is sometimes called the Michaelis complex. The enzyme then catalyzes the chemical step in the reaction and releases the product. The kinetics of many enzymes is adequately described by the simple Michaelis-Menten model, but all enzymes have internal motions that are not accounted for in the model and can have significant contributions to the overall reaction kinetics. This can be modeled by introducing several Michaelis-Menten pathways that are connected with fluctuating rates, which is a mathematical extension of the basic Michaelis Menten mechanism.

Software

ENZO

ENZO (Enzyme Kinetics) is a graphical interface tool for building kinetic models of enzyme catalyzed reactions. ENZO automatically generates the corresponding differential equations from a stipulated enzyme reaction scheme. These differential equations are processed by a numerical solver and a regression algorithm which fits the coefficients of differential equations to experimentally observed time course curves. ENZO allows rapid evaluation of rival reaction schemes and can be used for routine tests in enzyme kinetics.

Equilibrium Binding

A simple enzymatic reaction involves binding of free enzyme (E) to the ligand (A) to form the complex (EA). The reaction is reversible and hence law of mass action can be used to define the binding event with two terms namely association constant (K_a) and dissociation constant (K_d).

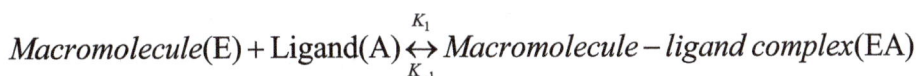

$$Macromolecule(\text{E}) + \text{Ligand(A)} \underset{K_{-1}}{\overset{K_1}{\leftrightarrow}} Macromolecule-ligand\ complex(\text{EA})$$

Where,

$$K_a = \frac{K_1}{K_{-1}} = \frac{[EA]}{[E][A]}$$

$$K_d = \frac{K_{-1}}{K_1} = \frac{[E][A]}{[EA]}$$

Both association and dissociation constants can be used to describe the equilibrium of the enzymatic reaction. In general, association constant K_a is used to explain the equilibrium of the reaction and dissociation constant K_d is used to explain the enzyme kinetics. The affinity of binding depends on the concentration of the substrate.

Association constant K_a is inversely proportional to the concentration of the substrate and hence more K_a will lead to more affinity. The dissociation constant K_d is directly proportional to concentration and hence lower the value of K_d, strong binding affinity will be observed.

Binding Curve

The relationship between the fraction of free ligand to the fraction of ligand bound to macromolecule can be arrived at from the expression of K_d and usually shown as a binding curve. Initially, only the total enzyme and ligand concentrations are known. So, from law of mass conservation,

$$[E]_0 = [E] + [EA] \quad and$$
$$[A]_0 = [A] + [EA]$$

Concentration of bound ligand can be determined through experiments. When a ligand has only one binding site for an enzyme, its concentration in bound form will be,

$$[A]_{bound} = \frac{[E_0][A]}{K_d + [A]}$$

However, in real conditions, most of the enzymes have more than one binding site for a specific ligand. For e.g., Haemoglobin (Hb) has 4 distinct binding sites for oxygen molecule.

Haemoglobin present in our blood has the ability to bind oxygen and it carries it from lungs to the tissues. The partial pressure of oxygen plays important role in binding of oxygen to haemoglobin. Also, the concentration of Haemoglobin present in blood is essential to determine its binding efficiency. When all the binding sites in Hb molecules are occupied, the blood is said to be 100% saturated and cannot carry any more oxygen. Under the different partial pressures of oxygen the saturation level of haemoglobin is influenced (Figure below).

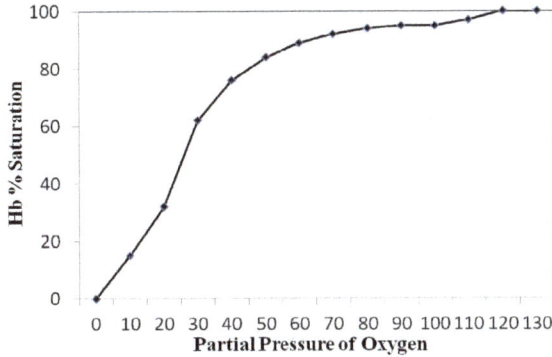

Binding curve for Haemoglobin showing the %saturation at different partial pressures of oxygen

In certain situations, one of the reactant might be available at huge amount and hence we can neglect the change in concentration of that reactant throughout the reaction. Consider any hydrolysis reaction, where hydrogen ions will be present in excess amount, making it difficult to detect any change in its concentration. In these situations, the K_d for excess reactant is not considered and an apparent K_d will be calculated which is concentration of the reactant times K_d.

In biological processes, the genetic material either DNA/RNA, requires proper binding with specific proteins for their active mechanisms like transcription and translation to be executed. RNA polymerase enzyme complex should bind to DNA efficiently to initiate and proceed with transcription. In addition, binding of small molecules such as transcription factors, activators/repressors to this enzyme complex is crucial for the process to take over. This will give a clear idea on the importance of binding.

Complexity in Binding

Binding of a single protein molecule to DNA is complex and is difficult to understand while the binding of multiple protein complexes that are non-equivalent makes the binding process much more complex (Figure below).

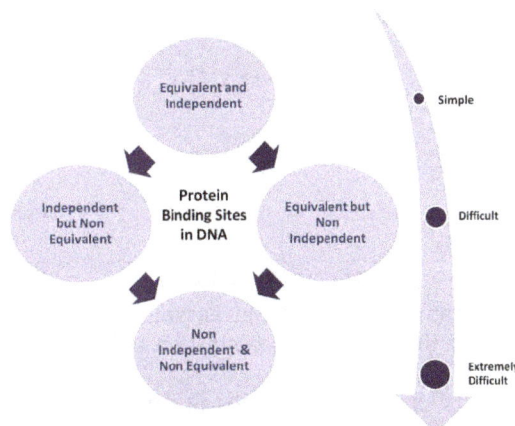

Complexity in analysis of protein-DNA binding at equilibrium

It is essential to optimise the conditions for effective binding between DNA and protein. To achieve this, we should concentrate on quantitative analysis on the binding affinity and the occupancy of the binding site (either fully or partially occupied); number of protein molecules binding to each DNA molecule; equilibrium binding constants and specificity of the binding; Cooperativity of binding in case of multiple proteins binding to same DNA molecule, etc.,

Cooperativity and Allostery

In most biological events the enzyme has the ability to bind another substrate molecule in a site different from that of the active site where the first substrate has been bound. In such conditions, binding of first substrate might have some influence over the binding of second substrate/ligand at a different site of the enzyme, where this indirect influence at a different site is called Co-operativity. The binding of another substrate at a different site apart from its active site is called allostery and enzymes showing this property are termed allosteric enzymes (allo – other).

Certain molecules or ligands that bind to the binding site can either be activators or inhibitors, depending on how they influence the binding at the second site when bound to the enzyme. Cooperativity of the ligands towards the enzyme can be either positive where the binding of the first ligand facilitates binding of other or negative where the condition is reverse. Whenever the number of binding sites in the enzyme is limited, the affinity of ligand is critical to determine the binding.

So far, we have seen the advantages of an enzymatic reaction and the steps involved, equilibrium of the binding, its complexity and the cooperative behaviour in enzymes. We shall discuss the kinetics of a simple single substrate enzymatic reaction.

Michaelis–Menten Kinetics

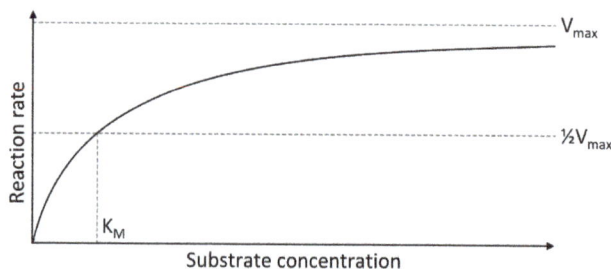

Michaelis–Menten saturation curve for an enzyme reaction showing the relation between the substrate concentration and reaction rate.

In biochemistry, '*Michaelis–Menten*' *kinetics* is one of the best-known models of enzyme kinetics. It is named after German biochemist Leonor Michaelis and Canadian physician Maud Menten. The model takes the form of an equation describing the rate of enzymatic reactions, by relating reaction rate v to $[S]$, the concentration of a substrate S. Its formula is given by

$$v = \frac{d[P]}{dt} = \frac{V_{max}[S]}{K_M + [S]}.$$

This equation is called the Michaelis–Menten equation. Here, V_{max} represents the maximum rate achieved by the system, at saturating substrate concentration. The Michaelis constant K_M is the substrate concentration at which the reaction rate is half of V_{max}. Biochemical reactions involving a single substrate are often assumed to follow Michaelis–Menten kinetics, without regard to the model's underlying assumptions.

Model

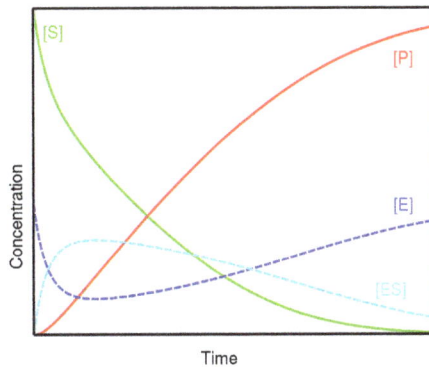

Change in concentrations over time for enzyme E, substrate S, complex ES and product P

In 1903, French physical chemist Victor Henri found that enzyme reactions were initiated by a bond (more generally, a binding interaction) between the enzyme and the substrate. His work was taken up by German biochemist Leonor Michaelis and Canadian physician Maud Menten, who investigated the kinetics of an enzymatic reaction mechanism, invertase, that catalyzes the hydrolysis of sucrose into glucose and fructose. In 1913, they proposed a mathematical model of the reaction. It involves an enzyme, E, binding to a substrate, S, to form a complex, ES, which in turn releases a product, P, regenerating the original enzyme. This may be represented schematically as

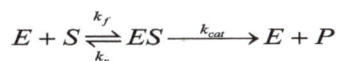

$$E + S \underset{k_r}{\overset{k_f}{\rightleftharpoons}} ES \xrightarrow{\; k_{cat} \;} E + P$$

where k_f (forward rate), k_r (reverse rate), and k_{cat} (catalytic rate) denote the rate constants, the double arrows between S (substrate) and ES (enzyme-substrate complex) represent the fact that enzyme-substrate binding is a reversible process, and the single forward arrow represents the formation of P (product).

Under certain assumptions – such as the enzyme concentration being much less than the substrate concentration – the rate of product formation is given by

$$v = \frac{d[P]}{dt} = V_{max} \frac{[S]}{K_M + [S]} = k_{cat}[E]_0 \frac{[S]}{K_M + [S]}.$$

The reaction order depends on the relative size of the two terms in the denominator. At low substrate concentration $[S] \ll K_M$ so that $v = k_{cat}[E]_0 \frac{[S]}{K_M}$. Under these conditions the reaction rate varies linearly with substrate concentration $[S]$ (first-order kinetics). However at higher $[S]$ with $[S] \gg K_M$, the reaction becomes independent of $[S]$ (zero-order kinetics) and asymptotically approaches its maximum rate $V_{max} = k_{cat}[E]_0$, where $[E]_0$ is the initial enzyme concentration. This rate is attained when all enzyme is bound to substrate. k_{cat}, the turnover number, is the maximum number of substrate molecules converted to product per enzyme molecule per second. Further addition of substrate does not increase the rate which is said to be saturated.

The Michaelis constant K_M is the $[S]$ at which the reaction rate is at half-maximum, and is an inverse measure of the substrate's affinity for the enzyme—as a small K_M indicates high affinity, meaning that the rate will approach V_{max} with lower $[S]$ than those reactions with a larger K_M. The value of K_M is dependent on both the enzyme and the substrate, as well as conditions such as temperature and pH.

The model is used in a variety of biochemical situations other than enzyme-substrate interaction, including antigen-antibody binding, DNA-DNA hybridization, and protein-protein interaction. It can be used to characterise a generic biochemical reaction, in the same way that the Langmuir equation can be used to model generic adsorption of biomolecular species. When an empirical equation of this form is applied to microbial growth, it is sometimes called a Monod equation.

Applications

Parameter values vary widely between enzymes:

Enzyme	K_M (M)	k_{cat} (1/s)	k_{cat}/K_M (1/(M*s))
Chymotrypsin	1.5×10^{-2}	0.14	9.3
Pepsin	3.0×10^{-4}	0.50	1.7×10^3
Tyrosyl-tRNA synthetase	9.0×10^{-4}	7.6	8.4×10^3
Ribonuclease	7.9×10^{-3}	7.9×10^2	1.0×10^5
Carbonic anhydrase	2.6×10^{-2}	4.0×10^5	1.5×10^7
Fumarase	5.0×10^{-6}	8.0×10^2	1.6×10^8

The constant k_{cat}/K_M (catalytic efficiency) is a measure of how efficiently an enzyme converts a substrate into product. Diffusion limited enzymes, such as fumarase, work at the theoretical upper limit of $10^8 - 10^{10}$ /M*s, limited by diffusion of substrate into the active site.

Michaelis–Menten kinetics have also been applied to a variety of spheres outside of biochemical reactions, including alveolar clearance of dusts, the richness of species pools, clearance of blood alcohol, the photosynthesis-irradiance relationship, and bacterial phage infection.

Derivation

Applying the law of mass action, which states that the rate of a reaction is proportional to the product of the concentrations of the reactants (i.e.[E][S]), gives a system of four non-linear ordinary differential equations that define the rate of change of reactants with time t

$$\frac{d[E]}{dt} = -k_f[E][S] + k_r[ES] + k_{cat}[ES]$$

$$\frac{d[S]}{dt} = -k_f[E][S] + k_r[ES]$$

$$\frac{d[ES]}{dt} = k_f[E][S] - k_r[ES] - k_{cat}[ES]$$

$$\frac{d[P]}{dt} = k_{cat}[ES].$$

In this mechanism, the enzyme E is a catalyst, which only facilitates the reaction, so that its total concentration, free plus combined, $[E]+[ES]=[E]_0$ is a constant. This conservation law can also be observed by adding the first and third equations above.

Equilibrium approximation

In their original analysis, Michaelis and Menten assumed that the substrate is in instantaneous chemical equilibrium with the complex, which implies

$$k_f[E][S] = k_r[ES].$$

From the enzyme conservation law, we obtain

$$[E] = [E]_0 - [ES].$$

Combining the two expressions above, gives us

$$k_f([E]_0 - [ES])[S] = k_r[ES].$$

Upon simplification, we get

$$[ES] = \frac{[E]_0[S]}{K_d + [S]}$$

where $K_d = k_r / k_f$ is the dissociation constant for the enzyme-substrate complex. Hence the velocity v of the reaction – the rate at which P is formed – is

$$v = \frac{d[P]}{dt} = \frac{V_{max}[S]}{K_d + [S]}$$

where $V_{max} = k_{cat}[E]_0$ is the maximum reaction velocity.

Quasi-steady-state approximation

An alternative analysis of the system was undertaken by British botanist G. E. Briggs and British geneticist J. B. S. Haldane in 1925. They assumed that the concentration of the intermediate complex does not change on the time-scale of product formation – known as the quasi-steady-state assumption or pseudo-steady-state-hypothesis. Mathematically, this assumption means $k_f[E][S] = k_r[ES] + k_{cat}[ES]$. Combining this relationship with the enzyme conservation law, the concentration of the complex is

$$[ES] = \frac{[E]_0[S]}{K_M + [S]}$$

where

$$K_M = \frac{k_r + k_{cat}}{k_f}$$

is known as the Michaelis constant, where k_r, k_{cat}, and k_f are, respectively, the constants for substrate unbinding, conversion to product, and binding to the enzyme. Hence the velocity v of the reaction is

$$v = \frac{d[P]}{dt} = \frac{V_{max}[S]}{K_M + [S]}.$$

Assumptions and Limitations

The first step in the derivation applies the law of mass action, which is reliant on free diffusion. However, in the environment of a living cell where there is a high concentration of proteins, the cytoplasm often behaves more like a gel than a liquid, limiting molecular movements and altering reaction rates. Although the law of mass action can be valid in heterogeneous environments, it is more appropriate to model the cytoplasm as a fractal, in order to capture its limited-mobility kinetics.

The resulting reaction rates predicted by the two approaches are similar, with the only difference being that the equilibrium approximation defines the constant as K_d, whilst the quasi-steady-state approximation uses K_M. However, each approach is founded upon a different assumption. The Michaelis–Menten equilibrium analysis is valid if the substrate reaches equilibrium on a much faster time-scale than the product is formed or, more precisely, that

$$\epsilon_d = \frac{k_{cat}}{k_r} \ll 1.$$

By contrast, the Briggs–Haldane quasi-steady-state analysis is valid if

$$\epsilon_m = \frac{[E]_0}{[S]_0 + K_M} \ll 1.$$

Thus it holds if the enzyme concentration is much less than the substrate concentration. Even if this is not satisfied, the approximation is valid if K_M is large.

In both the Michaelis–Menten and Briggs–Haldane analyses, the quality of the approximation improves as ϵ decreases. However, in model building, Michaelis–Menten kinetics are often invoked without regard to the underlying assumptions.

It is also important to remember that, while irreversibility is a necessary simplification in order to yield a tractable analytic solution, in the general case product formation is not in fact irreversible. The enzyme reaction is more correctly described as

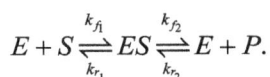

$$E + S \underset{k_{r_1}}{\overset{k_{f_1}}{\rightleftharpoons}} ES \underset{k_{r_2}}{\overset{k_{f_2}}{\rightleftharpoons}} E + P.$$

In general, the assumption of irreversibility is a good one in situations where one of the below is true:

1. The concentration of substrate(s) is very much larger than the concentration of products:

$$[S] \gg [P].$$

This is true under standard *in vitro* assay conditions, and is true for many *in vivo* biological reactions, particularly where the product is continually removed by a subsequent reaction.

2. The energy released in the reaction is very large, that is

$$\Delta G \ll 0.$$

In situations where neither of these two conditions hold (that is, the reaction is low energy and a substantial pool of product(s) exists), the Michaelis–Menten equation breaks down, and more complex modelling approaches explicitly taking the forward and reverse reactions into account must be taken to understand the enzyme biology.

Determination of Constants

The typical method for determining the constants V_{max} and K_M involves running a series of enzyme assays at varying substrate concentrations $[S]$, and measuring the initial reaction rate v_0. 'Initial' here is taken to mean that the reaction rate is measured after a relatively short time period, during which it is assumed that the enzyme-substrate complex has formed, but that the substrate concentration held approximately constant, and so the equilibrium or quasi-steady-state approximation remain valid. By plotting reaction rate against concentration, and using nonlinear regression of the Michaelis–Menten equation, the parameters may be obtained.

Before computing facilities to perform nonlinear regression became available, graphical methods involving linearisation of the equation were used. A number of these were proposed, including the Eadie–Hofstee diagram, Hanes–Woolf plot and Lineweaver–Burk plot; of these, the Hanes–Woolf plot is the most accurate. However, while useful for visualization, all three methods distort the error structure of the data and are inferior to nonlinear regression. Nonetheless, their use can still be found in modern literature.

In 1997 Santiago Schnell and Claudio Mendoza suggested a closed form solution for the time course kinetics analysis of the Michaelis–Menten kinetics based on the solution of the Lambert W function. Namely:

$$\frac{[S]}{K_M} = W(F(t))$$

where W is the Lambert W function and

$$F(t) = \frac{[S]_0}{K_M} \exp\left(\frac{[S]_0}{K_M} - \frac{V_{max}}{K_M} t \right).$$

The above equation has been used to estimate V_{max} and K_M from time course data.

Role of Substrate Unbinding

The Michaelis-Menten equation has been used to predict the rate of product formation in enzymatic reactions for more than a century. Specifically, it states that the rate of an enzymatic reaction will increase as substrate concentration increases, and that increased unbinding of enzyme-substrate complexes will decrease the reaction rate. While the first prediction is well established, the second has never been tested experimentally. To determine whether an increased rate of unbinding does in fact decrease the reaction rate, Shlomi Reuveni *et al.* mathematically analyzed the effect of enzyme-substrate unbinding on enzymatic reactions at the single-molecule level. According to the study, unbinding of an enzyme from a substrate can reduce the rate of product formation under some conditions, but may also have the opposite effect. As substrate concentrations increase, a tipping point can be reached where an increase in the unbinding rate results in an increase, rather than a decrease, of the reaction rate. The results indicate that enzymatic reactions can behave in ways that violate the classical Michaelis-Menten equation, and that the role of unbinding in enzymatic catalysis still remains to be determined experimentally.

Enzyme Inhibition

Enzyme inhibition can happen when the inhibitors (structural analogs of substrate) binds to the active site of the enzyme and prevents the catalysis. Enzyme inhibition is specific and is different from the alteration of structure of enzyme and reduction of reaction rate by environmental factors such as pH, temperature, presence of hydrophobic compounds, detergents etc., which are non specific. For e.g. consider sudden addition of an acid or base to the reaction mix which changes the pH and thereby influences the rate of the reaction. These are often confused with inhibition as they also reduce the turnover rate of enzymes.

In general, the binding of enzyme to the inhibitor is reversible but few of them bind covalently and become irreversible. The reversible and irreversible inhibitors have different kinetics. Michaelis-Menten kinetics explains the inhibition of enzyme in a single substrate complex, but the complexity increases with the number of substrates. The inhibitors also depend on their homology with the substrate apart from the nature of binding site and binding affinity.

Some inhibitors even bind stronger than the natural substrate because of specific interactions and act as antagonists. Most therapeutic drug molecules act in this way. Yet there are different forms of inhibition based on the affinity of inhibitor to the enzyme and substrate.

Competitive Inhibition

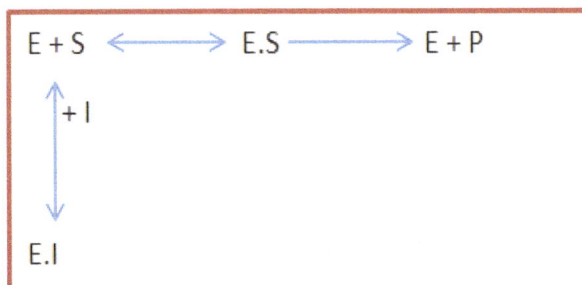

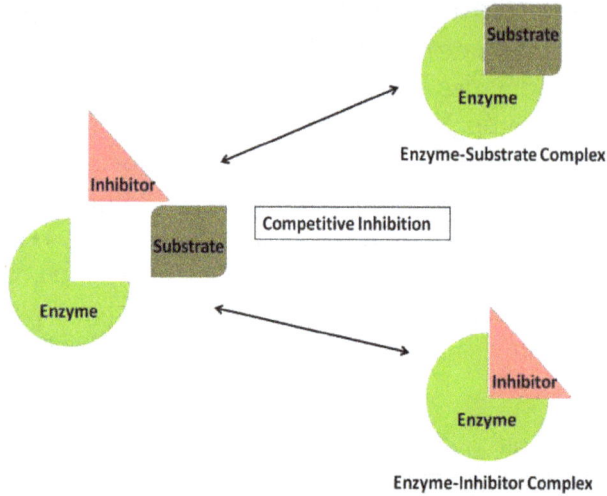

Schematic representation of competitive inhibition

Competitive inhibition occurs when the inhibitor is highly homologous to the substrate molecule and competes with substrate to bind to the free enzyme. So, either one of them can bind with an enzyme and not both together (Figure above). In this condition, there is a need for excess substrate to overcome the competition with inhibitor. Classical example for competitive inhibition is the molecule methotrexate which inhibits the action of dihydrofolate reductase to convert dihydrofolate to tetrahydrofolate.

The rate of product formation in the reaction is given by,

$$\frac{d[P]}{dt} = K_2[ES]$$

And from the Michaelis-Menten kinetics,

$$K_m = \frac{[E][S]}{[ES]}$$

$$[ES] = \frac{[E][S]}{K_m}$$

Similarly, rate of formation of enzyme – inhibitor complex will give,

$$K_I = \frac{[E][I]}{[EI]}$$

$$[EI] = \frac{[E][I]}{K_I}$$

The total enzyme concentration in the system will be the sum of the concentration of three forms in which the enzymes exists:

$$E_0 = [E] + [ES] + [EI]$$

$$E_0 = [E](1 + \frac{[S]}{K_m} + \frac{[I]}{K_I})$$

$$[ES] = \frac{[S]}{K_m}(\frac{E_0}{(1 + \frac{[S]}{K_m} + \frac{[I]}{K_I})})$$

$$\frac{d[P]}{dt} = V = K_2 \frac{[S]}{K_m}(\frac{E_0}{(1 + \frac{[S]}{K_m} + \frac{[I]}{K_I})})$$

$$K_m^{app} = 1 + \frac{[I]}{K_I}$$

$$V = \frac{V_{max}[S]}{K_m^{app} + [S]}$$

Hence in competitive inhibition, only the K_m is influenced and not the maximum velocity (V_{max}).

Uncompetitive Inhibition

Uncompetitive inhibition occurs when the inhibitor does not bind to the free enzyme and instead binds to the already formed enzyme substrate complex and makes the complex inactive (Figure below). This phenomenon of inhibition is commonly observed in multimeric enzymes.

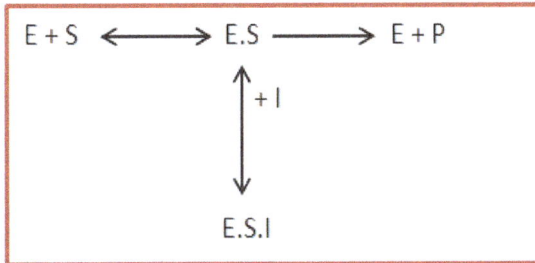

The rate of product formation in the reaction is given by,

$$\frac{d[P]}{dt} = K_2[ES]$$

And from the Michaelis-Menten kinetics,

$$K_m = \frac{[E][S]}{[ES]}$$

$$[ES] = \frac{[E][S]}{K_m}$$

Similarly, rate of formation of enzyme-substrate-inhibitor complex will give,

$$[ESI] = \frac{[ES][I]}{K_I}$$

$$[ESI] = \frac{[E][S][I]}{K_I K_m}$$

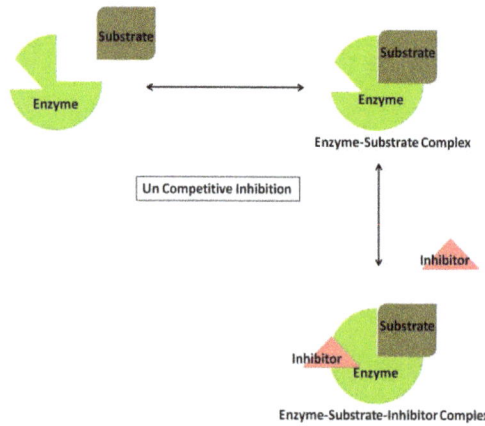

Schematic representation of uncompetitive inhibition

The total enzyme concentration in the system will be the sum of the concentration of three forms in which the enzymes exists:

$$E_0 = [E] + [ES] + [ESI]$$

$$E_0 = [E](1 + \frac{[S]}{K_m} + \frac{[I][S]}{K_I K_m})$$

$$[E] = (\frac{E_0}{1 + \frac{[S]}{K_m}(1 + \frac{[I]}{K_I})})$$

$$[ES] = \frac{[S]}{K_m}(\frac{E_0}{(1 + \frac{[S]}{K_m} + \frac{[I]}{K_I})})$$

$$\frac{d[P]}{dt} = V = \frac{K_2 E_0 [S]}{K_m + [S](1 + \frac{[I]}{K_I})}$$

$$V_m^{app} = \frac{K_2 E_0}{(1 + \frac{[I]}{K_I})}$$

$$K_m^{app} = \frac{K_m}{(1 + \frac{[I]}{K_I})}$$

$$V = \frac{V_m^{app}[S]}{K_m^{app} + [S]}$$

In uncompetitive inhibition, both K_m as well as maximum velocity (V_{max}) is influenced.

Non-Competitive Inhibition

Non-Competitive inhibition is where; the inhibitor binds to the different site in the enzyme. So, in contrast to competitive inhibition, they can bind along with substrate to the enzyme and here both EI and ESI is inactive Figure.

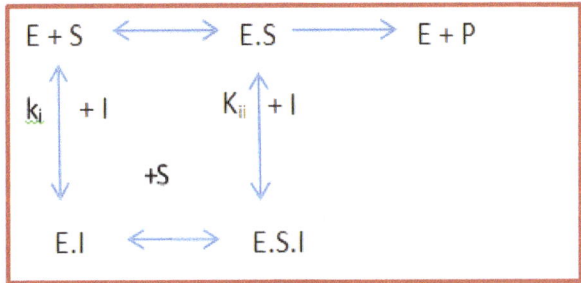

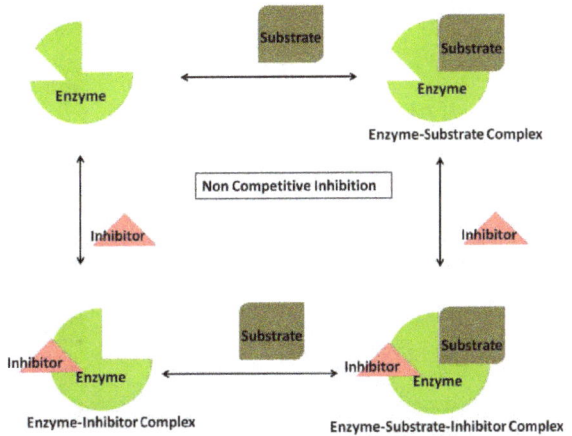

Schematic representation of non competitive inhibition

The rate of product formation in the reaction is given by,

$$\frac{d[P]}{dt} = K_2[ES]$$

And from the Michaelis-Menten kinetics,

$$K_m = \frac{[E][S]}{[ES]}$$

$$[ES] = \frac{[E][S]}{K_m}$$

Similarly, rate of formation of enzyme-substrate-inhibitor complex will give,

$$[ESI] = \frac{[ES][I]}{K_I}$$

$$[ESI] = \frac{[E][S][I]}{K_I K_m}$$

The total enzyme concentration in the system will be the sum of the concentration of four forms in which the enzymes exists:

$$E_0 = [E] + [ES] + [ESI]$$

$$E_0 = [E] + (1 + \frac{[S]}{K_m} + \frac{[I][S]}{K_I K_m} + \frac{[I]}{K_I})$$

$$[E] = (\frac{E_0}{(1 + \frac{[S]}{K_m})(1 + \frac{[I]}{K_I})})$$

$$[ES] = \frac{[S]}{K_m}(\frac{E_0}{(1 + \frac{[S]}{K_m})(1 + \frac{[I]}{K_I})})$$

$$\frac{d[P]}{dt} = V = \frac{K_2 E_0 [S]}{(K_m + [S])(1 + \frac{[I]}{K_I})}$$

$$V_m^{app} = \frac{K_2 E_0}{(1 + \frac{[I]}{K_I})}$$

$$V = \frac{V_m^{app}[S]}{K_m + [S]}$$

Non competitive inhibition influences the maximum velocity while the K_m does not changes. V_{max} is changed because high substrate concentration cannot prevent the binding of inhibitor.

Mixed Inhibition

Another mode of inhibition which is similar to that of non competitive inhibition but with an active ESI complex is termed the mixed inhibition. Such inhibition is common in metabolic feedback pathways. Enzymes showing this form of inhibition are generally allosteric in nature.

Identical Independent Binding Sites

Certain enzymes have multiple subunits each with identical binding sites. Yet there can be difference in the nature of binding sites, i.e., number of identical binding sites and the number of identical subunits of an enzyme.

In such cases, the binding follows simple successive steps, where the first ligand binds to the enzyme, then second ligand binds, and so on.

Consider a simple case where the enzyme 'E' has 'n' number of identical binding sites for the substrate 'A'. The binding event is as follows:

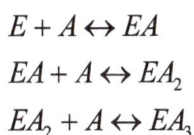

$$E + A \leftrightarrow EA$$
$$EA + A \leftrightarrow EA_2$$
$$EA_2 + A \leftrightarrow EA_3$$

And it goes on till 'n' sites are occupied.

$$EA_{n-1} + A \leftrightarrow EA_n$$

The corresponding expressions for macroscopic rate constants for the individual binding events can be given as:

$$K_1' = \frac{[E][A]}{[EA]}$$

$$K_2' = \frac{[EA][A]}{[EA_2]}$$

$$K_3' = \frac{[EA_2][A]}{[EA_3]}$$

And for the 'n' site,

$$K_1' = \frac{[EA_{n-1}][A]}{[EA_n]}$$

Imagine only there are 3 binding sites and thus the binding of 'A' to site 1 will lead to EA_1, site 2 will be EA_2 and site 3 will be EA_3. The individual microscopic rate constants for the first binding event will now become:

$$K_1 = \frac{[E][A]}{[E^A]}$$

$$K_2 = \frac{[E][A]}{[E_A]}$$

$$K_3 = \frac{[E][A]}{[EA]}$$

and hence we can arrive at

$$K_1' = \frac{1}{\left(\dfrac{1}{K_1} + \dfrac{1}{K_2} + \dfrac{1}{K_3}\right)}$$

If the binding sites are identical, $K_1 = K_2 = K_3 = K$

and hence

$$K_1' = \frac{K}{3}$$

Similarly for second binding if we substitute the microscopic rate constants,

$$K_2' = \frac{K_{13}K_{21}K_{23} + K_{12}K_{13}K_{23} + K_{13}K_{21}K_{32}}{K_{13}K_{23} + K_{12}K_{23} + K_{13}K_{21}}$$

where,

$$K_{12} = \frac{[E^A][A]}{E_A^A}$$

Similarly make the expression for all the microscopic constants. Here, when all binding sites are identical, all the microscopic constants are equal to K and hence,

$$K_2' = K$$

Similarly for the third binding step,

$$K_{123} = \frac{[AE^A][A]}{AE_A^A}$$

and applying the condition of identical binding sites,

$$K_3' = K_{123} + K_{132} + K_{321}$$

$$K_3' = 3K$$

We can generalise the relationship between the macroscopic and microscopic rate constants for 'n' binding sites as:

$$K'_d = K_d \frac{i}{n-i+1}$$

The number of possible orientations given by 'Ω' is also dependent on 'i'

$$\Omega = \frac{n!}{(n-i)!\,i!}$$

A saturation function 'r' is generally defined to make the binding equation simple. It is the quotient from the ratio of ligand bound to the enzyme to the total concentration of the macromolecule/enzyme.

$$r = \frac{[A]_{bound}}{[E]_0} = \frac{[EA]+2[EA_2]+3[EA_3]+.....+n[EA_n]}{[E]+[EA]+[EA_2]+[EA_3]+...+[EA_n]}$$

Substituting the following in r from the macroscopic rate expressions,

$$[EA] = \frac{[E][A]}{K'_1}$$

$$[EA_2] = \frac{[E][A]^2}{K'_1 K'_2}$$

$$[EA_3] = \frac{[E][A]^3}{K'_1 K'_2 K'_3}$$

Adair's Equation=>

$$r = \frac{\dfrac{[A]}{K'_1} + \dfrac{2[A]^2}{K'_1 K'_2} + \dfrac{3[A]^3}{K'_1 K'_2 K'_3} + ... + \dfrac{n[A]^n}{K'_1 K'_2 K'_3 ... K'_n}}{1 + \dfrac{[A]}{K'_1} + \dfrac{[A]^2}{K'_1 K'_2} + \dfrac{[A]^3}{K'_1 K'_2 K'_3} + ... + \dfrac{[A]^n}{K'_1 K'_2 K'_3 ... K'_n}}$$

$$r = \frac{\sum_{i=1}^{n} \dfrac{i[A]^i}{\prod_{j=1}^{i} K'_j}}{1 + \sum_{i=1}^{n} \dfrac{i[A]^i}{\prod_{j=1}^{i} K'_j}}$$

$$r = \frac{\sum_{i=1}^{n} i(\prod_{j=1}^{i} \frac{n-j+1}{j})(\frac{[A]^i}{K_d})}{1 + \sum_{i=1}^{n} (\prod_{j=1}^{i} \frac{n-j+1}{j})(\frac{[A]^i}{K_d})}$$

Using the binomial function,

$$\binom{n}{i} = \left(\frac{n!}{i!(n-1)!}\right)$$

$$r = \frac{\sum_{i=1}^{n} i\binom{n}{i}(\frac{[A]}{K_d})^i}{1+\sum_{i=1}^{n}\binom{n}{i}(\frac{[A]}{K_d})^i}$$

On application of binomial rule, the equation can be converted as:

$$r = \frac{n\left(\frac{[A]}{K_d}\right)(1+\frac{[A]}{K_d})^{n-1}}{(1+\frac{[A]}{K_d})^n}$$

This on simplification gives the general binding equation.

$$r = \frac{n[A]}{K_d +[A]}$$

This equation resembles the Michaelis-Menten kinetics and can be graphically represented as follows. This expression when plotted as a direct or Scatchard plot, gives a direct measure of the number of binding sites.

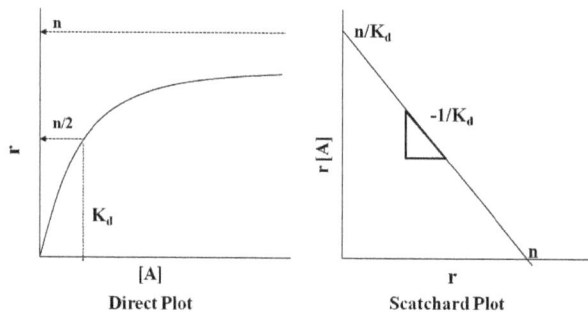

Direct plot and Scatchard plot giving information on the number of binding sites

When the concentration of macromolecule is not known, the saturation function (Ȳ) is reduced by 'n' and hence becomes,

$$\bar{Y} = \frac{[A]}{K_d +[A]}$$

Ȳ at saturation reaches the value 1 which gives the portion of ligand bound per binding site of the macromolecule.

Symmetry Model of Cooperativity

Jacques Monod and others presented a model to define the concept of cooperativity in case of allosteric enzymes. It is also referred as concerted model and it is based on few assumptions:

Allosteric enzyme is an oligomer composed of 'n' of identical protomers. Protomers occupy equal positions in the enzyme molecule and will possess atleast one symmetrical axis. Enzyme will exist in either of the two conformations, - tense(T) or relaxed(R) having diverse energy potential. No intermediate forms are present in the system. Symmetry is conserved in the reaction through transition from one form into the other. Tensed (T) state has low ligand affinity than the relaxed (R) state and has a low activity. This T state is favoured in the absence of ligand molecule.

Conformation states in Concerted Model of Cooperativity

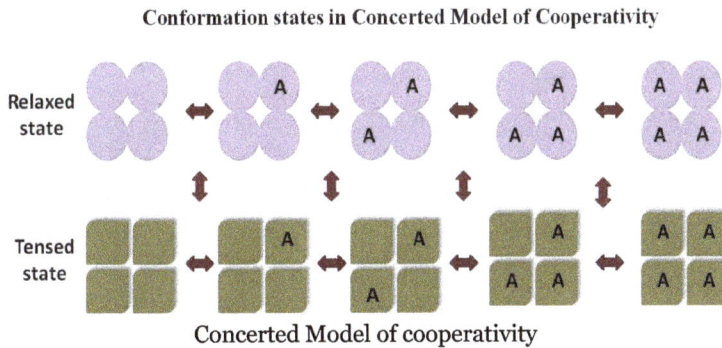

Concerted Model of cooperativity

If there is no deviation from these assumptions, sigmoidal saturation curves are obtained. In a linear plot, significant deviation from straight line will be observed in cases where the co-operativity becomes apparent. So, Hill plot can be used to study such systems as it will clearly distinguish the two types of co operativity. The straight lines in the plot represents the binding of ligand to the two distinct states of enzyme T and R, while the distance between two lines represent the energy difference among the tensed and relaxed states.

The maximum possible slope of the straight line is representative of the cooperative strength and is often referred as Hill Coefficient (n_H).

Initially, the first ligand binds to the high affinity R form and the R state is now shifted from equilibrium. To restore it, the molecule from excess T form is transformed into R form. Then, when the second ligand comes, both forms of enzyme are present in equilibrium and there will be further free sites in R state that are disturbed from equilibrium.

Significance of the Hill Coefficient

Effectively the Hill coefficient ranges between 1 and n. When all the macromolecules are shifted to the relaxed form, the value of n_H becomes 1. It describes the strength of cooperativity and if its value is in pair with the concentration of promoters, the cooperativity will be more evident. The Hill coefficient has a drawback that it does not directly state the number of subunits in the macromolecule.

Thus, we understand that in the concerted model of Cooperativity, the equilibrium of 2 forms of enzymes depend on the allosteric centers.

Non-Identical and Interacting Binding Sites

The next mode of binding relates to the condition when the binding sites are non-identical and interacting. This mode of binding is not generally found in biological systems. Difference in binding sites makes the binding pattern more different to that of identical interacting binding sites. Haemoglobin has two different binding sites but the sites have similar binding constants and hence were discussed under the identical binding site category. This can also be considered under the non-identical interacting category of binding.

Consider that there are two non-identical and interacting binding sites in a macromolecule. The macromolecule E will exist in four states, free form $[E_0]$ without any ligand [E], first site occupied $[E_1]$, second site occupied $[E_2]$, and two sites occupied [E_1']. The macroscopic rate constants for this condition can be written as:

$$K_1 = \frac{[E_1]}{[E_0][S]}$$

$$K_2 = \frac{[E_1']}{[E_0][S]}$$

$$K_3 = \frac{[E_2]}{[E_1][S]}$$

$$K_4 = \frac{[E_2]}{[E_1'][S]}$$

If we rewrite these equations, we arrive at

$$K_1 K_3 = K_2 K_4$$

For the non-identical interacting binding, the saturation function can be defined as:

$$\bar{Y} = \frac{1}{2} \frac{[E_1'] + [E_1] + 2[E_2]}{[E_0] + [E_1] + [E_1'] + [E_2]}$$

$$\bar{Y} = \frac{[E_0][S](K_1 + K_2) + 2[E_1][S]K_3}{[E_0] + [E_0][S](K_1 + K_2) + [E_1][S]K_3}$$

$$\bar{Y} = \frac{K_1[S] + K_2[S] + 2K_1 K_3[S]^2}{1 + K_1[S] + K_2[S] + K_1 K_3[S]^2}$$

If we define the following constants,

$$J = \frac{1}{2}(K_1 + K_2)$$

$$J' = \frac{2K_1K_3}{(K_1 + K_2)}$$

$$X' = J[S]$$

$$\beta' = \frac{J'}{J}$$

Then the saturation function becomes,

$$\bar{Y} = \frac{X'(1 + X'^1\beta')}{1 + 2X' + \beta'X'^2}$$

When $K_1 = K_2$, $X = X'$ and $\beta = \beta'$

When $K_1 = K_3$ and $K_2 = K_4$

$$\bar{Y} = \frac{K_1[S] + K_2[S] + 2K_1K_3[S]^2}{1 + K_1[S] + K_2[S] + K_1K_3[S]^2}$$

$$\bar{Y} = \frac{K_1[S]}{1 + K_1[S]} + \frac{K_2[S]}{1 + K_2[S]}$$

So, the expression for β' can be rewritten as,

$$\beta' = \frac{4K_1K_2}{4K_1K_2 + (K_1 - K_2)^2}$$

For non-identical binding sites, the value of β' is < 1. From this, we can clearly make out that the proteins having independent binding sites or interacting binding sites and the binding of second ligand or substrate molecule are being influenced by the first ligand.

When we consider the fact of co-existence of positive and negative cooperativity, the negative cooperativity shows a curvature similar to the non identical binding sites. This superimposition of negative cooperativity with that of the non-identical subunits can be correlated to the regulation of the binding events. But there are no models in real biological systems to explain this mode of binding.

Physiological Importance of Cooperativity and Allostery

Most biological regulatory events such as receptor- ligand binding and transport mechanisms follow Cooperativity. Cooperativity of the system is stable over the fluctuations in ligand concentrations. Allostery has an effect over the conformation of active sites. Cooperativity is advantageous in the fact that it gives a steep rise in the saturation curve in the middle. Here, in the sigmoidal curve, even small change in concentration changes the activity to a greater extent.

Models of Cooperativity do not explain fully the cooperativity in monomeric enzynes. We can correlate this with the binding of oxygen to monomeric myoglobin versus the binding of haemoglobin with four subunits. In reality, certain monomeric enzymes such as ribonucleases also show sigmoidal binding curves. Such mechanisms are time dependent and require extremely fast turnover and very slow transition from inactive to active state of enzyme. This phenomenon is called kinetic Cooperativity.

The strength of cooperativity can be easily determined by the nH of the Hill equation. Positive cooperativities have value of n_H ranging between 1 and n, while negative cooperativities have $n_H < 1$.

RS value is the ratio of concentration of ligand at 90% and 10% saturation. It is a direct measure of cooperativity for the hyperbolic systems, where the value is always 81. In case of positive cooperativity, the curve will be much steeper and hence RS value decreases. The basic difference between RS value and Hill coefficient is that, the Hill coefficient gives the cooperativity at specific point while the value of RS represents the same at different ligand concentrations. These are the important parameters to assess the cooperativity in a biological enzymatic reaction.

References

- Hammes G (2002). "Multiple conformational changes in enzyme catalysis". Biochemistry. 41 (26): 8221–8. doi:10.1021/bi0260839. PMID 12081470

- Beal, S. L. (1983). "Computation of the explicit solution to the Michaelis-Menten equation". Journal of Pharmacokinetics and Biopharmaceutics. 11 (6): 641–657. doi:10.1007/BF01059062. PMID 6689584

- Danson, Michael; Eisenthal, Robert (2002). Enzyme assays: a practical approach. Oxford [Oxfordshire]: Oxford University Press. ISBN 0-19-963820-9

- Xie XS, Lu HP (June 1999). "Single-molecule enzymology". J. Biol. Chem. 274 (23): 15967–70. doi:10.1074/jbc.274.23.15967. PMID 10347141

- Gibson QH (1969). "Rapid mixing: Stopped flow". Methods in Enzymology. Methods in Enzymology. 16: 187–228. doi:10.1016/S0076-6879(69)16009-7. ISBN 978-0-12-181873-9

- Lu H (2004). "Single-molecule spectroscopy studies of conformational change dynamics in enzymatic reactions". Current pharmaceutical biotechnology. 5 (3): 261–9. doi:10.2174/1389201043376887. PMID 15180547

- Baillie T, Rettenmeier A (1986). "Drug biotransformation: mechanistic studies with stable isotopes". Journal of clinical pharmacology. 26 (6): 448–51. doi:10.1002/j.1552-4604.1986. tb03556.x. PMID 3734135

- Duggleby RG (1995). "Analysis of enzyme progress curves by non-linear regression". Methods in Enzymology. Methods in Enzymology. 249: 61–90. doi:10.1016/0076-6879(95)49031-0. ISBN 978-0-12-182150-0. PMID 7791628

- Stroppolo ME, Falconi M, Caccuri AM, Desideri A (2001). "Superefficient enzymes". Cell. Mol. Life Sci. 58 (10): 1451–60. doi:10.1007/PL00000788. PMID 11693526

- Walsh, Ryan (2012). "Ch. 17. Alternative Perspectives of Enzyme Kinetic Modeling". In Ekinci,

Deniz. Medicinal Chemistry and Drug Design (PDF). InTech. pp. 357–371. ISBN 978-953-51-0513-8

- Briggs GE, Haldane JB (1925). "A Note on the Kinetics of Enzyme Action". The Biochemical Journal. 19 (2): 339–339. doi:10.1042/bj0190338. PMC 1259181. PMID 16743508

- Kraut J (1977). "Serine proteases: structure and mechanism of catalysis". Annu. Rev. Biochem. 46: 331–58. doi:10.1146/annurev.bi.46.070177.001555. PMID 332063

- Schnell, S.; Mendoza, C. (1997). "A closed form solution for time-dependent enzyme kinetics". Journal of Theoretical Biology. 187 (2): 207–212. doi:10.1006/jtbi.1997.0425

- Lehninger, A.L.; Nelson, D.L.; Cox, M.M. (2005). Lehninger principles of biochemistry. New York: W.H. Freeman. ISBN 978-0-7167-4339-2

- Ricard J, Cornish-Bowden A (July 1987). "Co-operative and allosteric enzymes: 20 years on". Eur. J. Biochem. 166 (2): 255–72. doi:10.1111/j.1432-1033.1987.tb13510.x. PMID 3301336

- Mathews, C.K.; van Holde, K.E.; Ahern, K.G. (10 Dec 1999). Biochemistry (3 ed.). Prentice Hall. ISBN 978-0-8053-3066-3

- Cleland WW (January 2005). "The use of isotope effects to determine enzyme mechanisms". Arch. Biochem. Biophys. 433 (1): 2–12. doi:10.1016/j.abb.2004.08.027. PMID 15581561

- Keener, J.; Sneyd, J. (2008). Mathematical Physiology: I: Cellular Physiology (2 ed.). Springer. ISBN 978-0-387-75846-6

- Northrop D (1981). "The expression of isotope effects on enzyme-catalyzed reactions". Annu. Rev. Biochem. 50: 103–31. doi:10.1146/annurev.bi.50.070181.000535. PMID 7023356

- Leskovac, V. (2003). Comprehensive enzyme kinetics. New York: Kluwer Academic/Plenum Pub. ISBN 978-0-306-46712-7

- Cleland WW (1982). "Use of isotope effects to elucidate enzyme mechanisms". CRC Crit. Rev. Biochem. 13 (4): 385–428. doi:10.3109/10409238209108715. PMID 6759038

- Kraut D, Carroll K, Herschlag D (2003). "Challenges in enzyme mechanism and energetics". Annu. Rev. Biochem. 72: 517–71. doi:10.1146/annurev.biochem.72.121801.161617. PMID 12704087

Permissions

All chapters in this book are published with permission under the Creative Commons Attribution Share Alike License or equivalent. Every chapter published in this book has been scrutinized by our experts. Their significance has been extensively debated. The topics covered herein carry significant information for a comprehensive understanding. They may even be implemented as practical applications or may be referred to as a beginning point for further studies.

We would like to thank the editorial team for lending their expertise to make the book truly unique. They have played a crucial role in the development of this book. Without their invaluable contributions this book wouldn't have been possible. They have made vital efforts to compile up to date information on the varied aspects of this subject to make this book a valuable addition to the collection of many professionals and students.

This book was conceptualized with the vision of imparting up-to-date and integrated information in this field. To ensure the same, a matchless editorial board was set up. Every individual on the board went through rigorous rounds of assessment to prove their worth. After which they invested a large part of their time researching and compiling the most relevant data for our readers.

The editorial board has been involved in producing this book since its inception. They have spent rigorous hours researching and exploring the diverse topics which have resulted in the successful publishing of this book. They have passed on their knowledge of decades through this book. To expedite this challenging task, the publisher supported the team at every step. A small team of assistant editors was also appointed to further simplify the editing procedure and attain best results for the readers.

Apart from the editorial board, the designing team has also invested a significant amount of their time in understanding the subject and creating the most relevant covers. They scrutinized every image to scout for the most suitable representation of the subject and create an appropriate cover for the book.

The publishing team has been an ardent support to the editorial, designing and production team. Their endless efforts to recruit the best for this project, has resulted in the accomplishment of this book. They are a veteran in the field of academics and their pool of knowledge is as vast as their experience in printing. Their expertise and guidance has proved useful at every step. Their uncompromising quality standards have made this book an exceptional effort. Their encouragement from time to time has been an inspiration for everyone.

The publisher and the editorial board hope that this book will prove to be a valuable piece of knowledge for students, practitioners and scholars across the globe.

Index

www.ingramcontent.com/pod-product-compliance
Lightning Source LLC
Chambersburg PA
CBHW061951190326
41458CB00009B/2846